Lecture Notes in Economics and Mathematical Systems

Stefan Schöne

Auctions in the Electricity Market

Bidding when Production Capacity Is Constrained

Stefan Schöne
Humboldt-Universität zu Berlin
School of Business and Economics
Institute of Finance
Spandauer Straße 1
10099 Berlin
Germany
schoene@wiwi.hu-berlin.de

Dissertation, Humboldt-Universität zu Berlin, 2008

ISBN 978-3-540-85364-0 e-ISBN 978-3-540-85365-7
DOI 10.1007/978-3-540-85365-7

Lecture Notes in Economics and Mathematical Systems ISSN 0075-8442

Library of Congress Control Number: 2008934457

Cover design: SPi Publishing Services

Printed on acid-free paper

9 8 7 6 5 4 3 2 1

springer.com

Acknowledgements

This work was written during my work as a research assistant at the Institute of Finance of the School of Business and Economics at the Humboldt-Universitaet zu Berlin. It is a result of a development process. I would like to thank all people who have supported and helped me.

Special thanks go to my academic teacher and Professor Dr. Sigrid Müller, who provided useful hints and ideas for my work. She accompanied and promoted the development with her encouragement, her willingness to discuss issues, and her constructive reviews. I would also like to thank everybody at the Institute of Finance, especially Dr. Matthias Pytlik, for the cooperative and creative atmosphere.

I am grateful to the "Hypo-Bank-Stifungsfonds zur Förderung bankwissenschaftlicher Nachwuchskräfte" for the opportunity to present and discuss my results in seminars for doctoral students of various universities.

My very special thanks go to my parents, to whom I dedicate this work.

Berlin, September 2008 *Stefan Schöne*

Contents

List of Figures

Acronyms

BDEW	BDEW Bundesverband der Energie- und Wasserwirtschaft e. V.
CEGB	Central Electricity Generating Board
e.g.	For example
EEX	EEX European Energy Exchange European Energy Exchange
Eidgen. Dep. UVEK	Eidgenössisches Departement für Umwelt, Verkehr, Energie und Kommunikation UVEK
EnWG	Energiewirtschaftsgesetz
Fn.	Footnote
GW	Gigawatt
i.e.	Id est
kWh	Kilowatt hour
LPX	LPX Leipzig Power Exchange
MW	Megawatt
NGC	National Grid Company PLC
p.	Page
pp.	Pages
RECs	Regional Electricity Companies
TWh	Terawatt hour
U.K.	United Kingdom
VDEW	Verband der Elektrizitätswirtschaft - VDEW - e. V.

Symbols

$\forall$	For all
$\mathbb{R}^{+}$	Set of all positive real numbers
α	Fraction of production capacity that the winner serves
β	Fraction of production capacity the loser serves
β_f	Fraction of winner's production capacity the loser can serve
γ	Auxiliary variable
λ	Loser's unused production capacity relative to demand
ξ	Auxiliary variable for integration
$\pi(b,c)$	Expected profit given bid b and marginal costs c
$\pi_d^*(c)$	Expected profit in a discriminatory auction
$\pi_d^{*i}(c)$	$\pi_d^*(c)$ with independently distributed marginal costs
$\pi_d^{u^{*i}}(c)$	$\pi_d^{u^*}(c)$ with independently distributed marginal costs
$\pi_d^{u^*}(c)$	Expected contribution margin per capacity unit in a discriminatory auction
$\pi_g^*(c)$	Expected profit in a generalized second-price auction
$\pi_g^{*i}(c)$	$\pi_g^*(c)$ with independently distributed marginal costs
$\pi_g^{u^{*i}}(c)$	$\pi_g^{u^*}(c)$ with independently distributed marginal costs
$\pi_g^{u^*}(c)$	Expected contribution margin per capacity unit in a generalized second-price auction
$\pi_u^*(c)$	Expected profit in a uniform-price auction
$\pi_u^{*i}(c)$	$\pi_u^*(c)$ with independently distributed marginal costs

$\pi_u^{u^{*i}}(c)$	$\pi_u^{u^*}(c)$ with independently distributed marginal costs
$\pi_u^{u^*}(c)$	Expected contribution margin per capacity unit in a uniform-price auction
τ	Auxiliary variable for integration
$1_{\{\text{condition}\}}$	Indicator function: $\begin{cases} 1 : \text{condition holds true,} \\ 0 : \text{otherwise} \end{cases}$
a	Constant showing the degree of affiliation of marginal costs in the example
$\hat{b}$	Upper limit of the bid function
$b(c)$	Bid function given marginal costs c
$b(c)'$	First derivative of bid function $b(c)$ with respect to marginal costs c
$b(c)''$	Second derivative of bid function $b(c)$ with respect to marginal costs c
$b^{-1}(\cdot)$	Inverse bid function with $b^{-1}(b(c)) = c$
$b^i(c)$	$b(c)$ with independently distributed marginal costs
$b_d^*(c)$	Optimal bid function of c in a discriminatory auction
$b_d^*(c)'$	First derivative of the optimal bid function in a discriminatory auction
$b_d^{*i}(c)$	$b_d^*(c)$ with independently distributed marginal costs
$b_g^*(c)$	Optimal bid function of c in a generalized second-price auction
$b_u^*(c)$	Optimal bid function of c in a uniform-price auction
$b_u^*(c)'$	First derivative of the optimal bid function in a uniform-price auction
$b_u^{*i}(c)$	$b_u^*(c)$ with independently distributed marginal costs
b_i	Bid of generator i
b^*	Optimal bidding strategy
$\bar{c}$	Upper limit of marginal costs
$\underline{c}$	Lower limit of marginal costs
c	Marginal costs
C_i	Information variable of generator i
c_i	Realized marginal costs of generator i

c^r	Realized production costs
d	Aggregated demand
$E[\cdot]$	Expectation operator
$E[c_d^i]$	$E[c_d]$ with independently distributed marginal costs
$E[c_d]$	Expected production costs per electricity unit in a discriminatory auction
$E[c_g^i]$	$E[c_g]$ with independently distributed marginal costs
$E[c_g]$	Expected production costs per electricity unit in a generalized second-price auction
$E[c_u^i]$	$E[c_u]$ with independently distributed marginal costs
$E[c_u]$	Expected production costs per electricity unit in a uniform-price auction
$E[c^i]$	$E[c]$ with independently distributed marginal costs
$E[c]$	Expected production costs per electricity unit
$F(c)$	Cumulative marginal distribution function of c
$f(c)$	Marginal distribution of c
$f(c_j, c_i)$	Probability density function of marginal costs c_i and c_j
$F(c_j \mid c_i)$	Conditional distribution function of c_j given c_i
$f(c_j \mid c_i)$	Conditional probability density function of c_j given c_i
$f_i(c_j, c_i)$	First partial derivative of $f(c_j, c_i)$ with respect to c_i
$F_i(c_j \mid c_i)$	First partial derivative of $F(c_j \mid c_i)$ with respect to c_i
$f_i(c_j \mid c_i)$	First partial derivative of $f(c_j \mid c_i)$ with respect to c_i
$f_j(c_j, c_i)$	First partial derivative of $f(c_j, c_i)$ with respect to c_j
$f_{ji}(c_j, c_i)$	Second partial derivative of $f(c_j, c_i)$ with respect to c_j and c_i
i	Index of the generator under review
j	Index of generator i's competitor
k	Production capacity of a generator
$\bar{p}$	Reserve price, price cap
$\bar{p}_d$	Upper limit of the reserve price in a discriminatory auction, which must not be reached
$\bar{p}_u$	Upper limit of the reserve price in a uniform-price auction, which must not be reached

p_d	Average price in a discriminatory auction
p_g	Average price in a generalized second-price auction
p_u	Uniform price (in a uniform-price auction)
p_d^i	p_d with independently distributed marginal costs
p_g^i	p_g with independently distributed marginal costs
p_u^i	p_u with independently distributed marginal costs
t	Auxiliary variable for integration
$y(c_j\|c_i)$	Auxiliary function (expected degree of production capacity utilization)
$y_i(c_j\|c_i)$	First derivative of $y(c_j\|c_i)$ with respect to c_i
$y_j(c_j\|c_i)$	First derivative of $y(c_j\|c_i)$ with respect to c_j
$y_{ji}(c_j\|c_i)$	Second derivative of $y(c_j\|c_i)$ with respect to c_j and c_i
$y_{jj}(c_j\|c_i)$	Second derivative of $y(c_j\|c_i)$ with respect to c_j

Chapter 1
Introduction

Abstract Investment opportunities can be assigned to asset classes such as stocks and bonds. Electricity is a new asset class, which emerged in the course of deregulation of electricity markets. The deregulation process started in the United Kingdom with the establishment of the power exchange in England and Wales in 1990. In 1998 Germany followed. Two power exchanges with different business approaches were initially established and merged after two years. The merged power exchange trades real electricity supply and derivatives using an auction system. The analysis of implemented auction rules based on auction theory helps to understand price movements of electricity. This may result in an improved trading environment with lower average prices for consumers.

Investors all over the world are looking for good investment opportunities. An advantage of exchange-traded assets is that they can be bought and sold all the time. Stocks and bonds are well-known examples. Nevertheless, other asset classes have attracted interest such as commodities like wheat, orange juice or metals. With the deregulation of electricity markets in many countries, electricity became a new commodity.

Trading electricity freely at power exchanges is more complex than trading other commodities because a blackout of the electricity grid can damage the economy of a large area. This work takes a look at the market microstructure of the electricity market. Power exchanges use auctions. Hence, three different auction rules are analyzed. The uniform-price auction is the most used one of them. This type is very common at stock exchanges because it is accepted to be fair.

Auction theory is used to analyze the electricity market because the behavior of all market participants can be modeled more precisely than with other approaches. The work presents numerical examples for illustration. Contrary to other literature, which covers independent production costs only, these examples are based on a simple distribution function, which allows for correlated production costs. Especially in electricity markets, a correlation of costs between producers is very likely, for example when fuel is bought at the world market.

S. Schöne, *Auctions in the Electricity Market*, Lecture Notes in Economics and Mathematical Systems 617,

The free trade of electricity is a rather new development in the history of the electricity industry. Traditionally, the electricity market was designed as a utility monopoly for a specific region. A particular company runs production facilities and the electricity grid. Having the control over the whole power grid is necessary to guarantee a stable power flow. Additionally, the installation of an electricity grid is very expensive. Both points were commonly accepted reasons to view the grid combined with power production as a natural monopoly. Whereas technical problems could be solved easily, the economic side of the business continually raised questions about the necessity of a monopoly. Governmental regulations and interventions were designed to help avoid market failures and provide consumers with fair prices.

The oil crises of the 1970s shed new light on the dependency of society on cheap energy.[1] Monopolies normally lead to higher consumer prices than competitive markets. Unfortunately, no methods could be found to transport electricity economically without using a grid. The necessity of a power grid makes the design of a competitive electricity market complicated because otherwise electricity could be sold like any other divisible and transportable product. But electricity is subject to some restrictions which have to be respected for the sake of the grid's safety.

The wholesale market for electricity in England and Wales, which was started in 1990, was organized as an auction. All large producers had to sell their entire electricity production in this auction. Small companies were allowed to participate.[2] The auction was designed to buy power at the lowest price and resell it through different providers to the consumers. The electricity grid was allocated to a newly established company acting as a system operator.[3] The restructured market became a reference model worldwide.[4]

Germany followed another approach. Before deregulation, the market was divided into regions. One utility, acting as a monopolist, controlled a region. Agreements regulated the power supply between the regions and the final access to the consumers. Contracts for direct access to consumers were necessary because local authorities normally owned that part of the electricity grid. Such agreements obviously inhibit competition, which required governmental regulation to mitigate resulting disadvantages.[5]

Deregulation started with a complete opening of the electricity market on April 29, 1998 when the Law of Reorganizing the Energy Industry Legislation became effective.[6] The key element was free access to the power grid for everybody without

---

[1] Gilbert, Kahn, and Newbery (1996) present some data about electricity prices from 1950 until 1990. Real prices fell in most countries until 1974. With the first oil price shock in 1973–74, this development was stopped and mostly reversed. See there on pp. 13 and 15.

[2] See Green and Newbery (1992), p. 929.

[3] A short market overview is given in Section 2.2. See also Beharrell (1991) and Vickers and Yarrow (1991), pp. 191–197.

[4] See Newbery (1998), p. 726.

[5] See Schiffer (1994), pp. 146–148 or Brunekreeft and Keller (2000), pp. 16 and 20–22.

[6] See Bundesgesetzblatt (1998).

discrimination. Utilities running power plants and owning parts of the electricity grid were obliged to separate both business parts organizationally. Outsourcing to different companies was not required.

The implemented regulations did not effectively enforce competition. A major problem was the negotiated grid access. Companies wanting to take part in the electricity market were competitors of the grid owner. The business separation did not solve the principal agent problems for free access.[7] Critics forced the government to amend the law. Production and sale of electricity is now more efficiently separated from running the power grid. Additionally, an authority was established to supervise the power grid and its access.[8]

In contrast to the market design in England and Wales, Germany did not install any specific market where producers could sell their electricity. Nevertheless, the industry was interested in an additional market place such as a power exchange which offers the chance to trade excess power or to buy electricity at a better price to match bilaterally agreed deliveries. Moreover, an exchange provides published prices, which can be used, for example, in agreements or for comparison reasons. Such public information can improve the allocation of capital and avoid the financial waste which often occurs in monopoly markets.

Four candidates applied to establish the first power exchange in German. An expert council set up by the government preferred the European Energy Exchange (EEX) in Frankfurt am Main.[9] Despite their recommendation, the LPX Leipzig Power Exchange (LPX) in Leipzig started its operations first in June 2000. The focus was the spot market, which was the trade of real electricity supply for each hour of the next business day. All bids for buying and selling electricity either of a single hour or specified blocks of hours were matched at one time and only once a day.[10] The EEX started its spot market in August 2000. Similar to the LPX, auctions for the 24 delivery periods with a length of one hour took place just once a day. Peak-load and baseload blocks, representing different delivery time periods, were traded in continuous trading with a discrete-time auction at the beginning and the end.[11] After a short while it became obvious that neither power exchange could attract enough business to survive alone. Hence, the LPX and EEX merged in 2002 into the EEX in Leipzig.[12] The business model of the LPX became the reference model for the new exchange.

The market share of the EEX was 6.4% of the annual electricity demand in Germany in 2002.[13] The exchange developed its business and entered new markets.

---

[7] A liberalized third party access to the power grid was also a feature of the Energy Act 1983 in the U.K. The reform was not very successful because it lacked protection for new entrants against anti-competitive behavior of the largest power producer. See Vickers and Yarrow (1991), p. 191.

[8] See EnWG (2005), Sections 6–10 and 29–35.

[9] See Müller (1999).

[10] See LPX GmbH (2000), pp. 5 and 10–11.

[11] See EEX AG (2000), pp. 4–5 and 7–8.

[12] See EEX AG and LPX GmbH (2001).

[13] 33.1 TWh of electricity was traded at the EEX in 2002. See EEX AG (2003). The electricity consumption of this year was 513.5 TWh in Germany. See VDEW (2003), p. 11.

Trading electricity with delivery in Austria started on April 1, 2005.[14] The first auction of electricity delivered in Switzerland took place on December 11, 2006.[15] The market share of EEX's spot market rose to 14.5% in 2006 and to approximately 18.6% in 2007.[16]

The revenue and income of the EEX have grown rapidly. This is not the result of trading in electricity alone. The trading portfolio now also contains coal, emission allowances and natural gas, the latter since 2007. Commodity prices of the EEX have reference character in Europe. This highlights the importance of this exchange.[17] Its European character will increase with the merger with the French energy exchange Powernext.[18]

The importance of power exchanges in daily business life raises questions about pricing rules. Especially in times when consumers are confronted with higher electricity prices, doubts arise about the fairness of these rules. Unfortunately, fuel prices have continued to rise for a long period of time, which will force producers to increase their prices. German consumers may face much higher costs as a result of decreasing production capacities due to the ban on new nuclear power plants.[19]Accusations over price manipulation are often an issue in times of rising prices. The EEX was a target of such accusations in 2006 and 2007, although they did not affect trading rules.[20] The electricity market in England and Wales presents a very different picture. It was redesigned from a uniform-price auction to a discriminatory pricing system[21] as a consequence of investigations into price manipulations. Criticism of the new market design remains.[22]

The EEX uses a uniform-price auction, the market in England and Wales a discriminatory auction. This raises questions about which auction should be used. The answer depends on the issue which is of greatest interest to the decision maker. Is the suppression of price manipulations a concern? Should the auction achieve production efficiency to save our resources? Or is only the lowest price for electricity to be sought?

This work focuses on the last two questions by analyzing the bidding behavior in both auctions and an adapted version of the Vickrey auction. Auction theory

---

[14] See EEX AG (2005).

[15] See EEX AG (2006).

[16] The amount of traded electricity in the spot market of the EEX was 88.7 TWh in 2006 and 123.7 TWh in 2007. See EEX AG (2008). In Germany and Austria, 607.6 TWh of electricity were consumed in 2006. The consumption in both countries and Switzerland for 2007 is estimated with 666.4 TWh. Because data for Austria and Switzerland in 2007 were not available, figures from 2006 are used instead. See BDEW (2008), STATISTIK AUSTRIA Bundesanstalt Statistik Österreich (2007), p. 107, and Eidgen. Dep. UVEK and Bundesamt für Energie BFE (2007), Table 1.

[17] See EEX AG (2007b), p. 3.

[18] See EEX AG (2007d).

[19] See exemplarily ap (2007) and dpa (2007).

[20] See EEX AG (2007c), manager-magazin.de (2007), stern.de (2007a), and stern.de (2007b).

[21] The market redesign motivated Federico and Rahman (2003) to analyze the new rules.

[22] See Harbord and McCoy (2000).

is used to model the market microstructure in the most precise way. To illustrate the results, examples are presented, which base on the same simple distribution function. The work is organized as follows: Chapter 2 reviews the literature on the electricity market. Chapter 3 is devoted to the market model based on auction theory. Market structure and trading rules are presented. Chapter 4 analyzes the game of generators. Examples are given. Chapter 5 concludes the analysis.

Chapter 2
Literature Review

Abstract This work focuses on bidding behavior and prices at power exchanges. For this a detailed knowledge of important issues of electricity markets is necessary. Section 2.1 gives a short insight into the electricity market. It also highlights the advantage of using auctions in power exchanges. Auctions are of great use balancing demand and supply of a whole market. The electricity market in England and Wales is the starting point of the model in this work. For this reason, Section 2.2 takes a look at this market. Sections 2.3 and 2.4 deal with approaches, which are often used in the literature analyzing bidding behavior at power exchanges. Critical aspects of these approaches are discussed. Section 2.5 gives an insight into literature primarily using auction theory.

2.1 Auctions in Electricity Markets

Consumers get their electricity out of sockets, which are connected with the electricity grid. This method of transmission is more economical than using storage batteries. But using an electricity grid makes the handling complicated. The reason is that electricity is a flow, which cannot be monitored perfectly. Due to Kirchhoff's Laws, it is impossible to determine which producer injected into the grid the electricity which was withdrawn by a consumer at the other end of the line. Additionally, the power flow must be balanced at every point in time to avoid damage to the grid and all connected devices.[1]

Electricity becomes tradable by creating property rights, which allow electricity to be injected or withdrawn at specific grid nodes. Balancing the electricity grid also requires information about injections and withdrawals as well as data concerning the capacity of the transmission lines. Exceeding these capacities could lead to a blackout. The system operator, who controls the electricity grid, has to respect additionally the loss of electricity that occurs during the transmission. Reasons for

[1] See, e.g. Wilson (2002), pp. 1300–1302.

S. Schöne, *Auctions in the Electricity Market*, Lecture Notes in Economics and Mathematical Systems 617,

the losses are resistance and magnetic induction. Resistance dissipates electrical energy into, e.g. heat. Magnetic induction occurs, for instance, when the voltage of transmission lines is changed in current transformers. Such changes are necessary because transmission at higher voltage is less lossy.[2]

An economic analysis of electricity markets must take into account two limiting factors. The first factor is the capacity of the transmission lines between different nodes in the electricity grid.[3] They must not be exceed otherwise a blackout is possible. The other limit is the aggregated amount of production capacity of electricity. It is technically impossible to produce more. Increasing transmission or production capacities is however time consuming and expensive.[4] This restricts market flexibility dramatically in the short term but leads to flexibility of supply in the long run.[5]

A crucial issue of the electricity market is the balancing of supply and demand or, which is the same, the injections into and withdrawals from the electricity grid. Only one agent should be responsible for this task, because a failure leads to tremendous damage to the economy in the case of a blackout. This person is often the system operator. He has to collect all necessary information without suppressing the competition between producers and consumers. An efficient way of doing this is using auctions to clear the market. Almost all electricity markets use uniform-price auctions, a specific type of auction.[6]

Auctions are of great use. They enforce competition among bidders, who can be either producers or consumers. Bidders in electricity auctions have usually private information about their utility functions. For a producer, it is the profit function. His profit is the result of the price, the amount of electricity sold in the auction and the costs to produce it. Costs are different for each producer and depend on the type of power plant. The efficiency of the power plants and the prices for fuel and other input factors are private information.

The auctioneer uses only submitted bids to determine the payments of each bidder. Hence, bidders incorporate their private information into the bids. The advantage is that no costly procedures must be implemented to guarantee truly submitted information. Additionally, the information of all market participants comes together at one place. The auctioneer has therefore all information to guarantee the safety of the electricity grid. The market can be cleared at lowest costs respecting all restrictions such as the capacity of transmission lines and the loss of electricity.

A uniform-price auction handles all nodes of the electricity grid equally. In cases of congested transmission lines, supply and demand from both sides of a congested line have to be managed differently. This causes a separation of the entire power

[2] Losses typically count for about 3–5% of the total power production. See Wu and Varaiya (1999), p. 80 and Chao and Peck (1996), p. 39.

[3] See e.g. Hogan (1998), pp. 5–8, where pricing rules for transmission service in competitive electricity markets are discussed.

[4] See Vickers and Yarrow (1991), pp. 189–190. They also give further economic characteristics of electricity supply.

[5] See Wilson (2002), p. 1301.

[6] See Cramton (2003), p. 5.

market and creates different electricity prices. The price difference can be seen as a transmission charge. It could be used as an incentive to expand the existing transmission lines. Such handling raises questions about how to insure against these differences. Solutions are either financial hedges or rights of physical access. The topology of the grid may require many separate markets. Reducing the number of separate markets for reasons of practicality induces false incentives, as has been seen in California.[7]

An electricity market usually consists of different markets. These markets result from different needs of the market participants: the producers, the consumers, and the system operator. Depending on legal restrictions, it may be allowed to run forward markets or day-ahead markets. They take place, for example, one day before the traded electricity is delivered. This gives the system operator the opportunity to detect possible congestion and reduce real-time operations in the spot market, which was discussed above.[8]

The safety of the electricity grid requires that enough production capacity is at hand in case of an unexpected falling out of a power plant. The system operator may use auctions to buy production capacity as reserves.[9] The market of reserves shows that the production capacity of a power plant can either be used to produce electricity directly for consumers or as an option to guarantee the safety of the grid. The various markets in an electricity market are all linked in some ways. This makes the analysis of the entire electricity market very complex. A reduction of the complexity by neglecting some issues is necessary to get answers to different questions.

2.2 England and Wales: The Reference Market

Many countries have deregulated or restructured their electricity markets since the 1990s. The main idea behind this process was the recognition that power production was not a natural monopoly. For a long period of time, the bundling of electricity production and power grid were seen as a natural monopoly[10] because balancing the electricity flow in the grid requires the production capacities of power plants. But a separation of production and transmission grid was feasible as well as the implementation of competition among electricity producers. Chile was a pioneer in this field, beginning the process as early as 1978.[11] Nevertheless, the deregulation process really started in the U.K. with the Electricity Act 1989. The prominent restructured market discussed in the literature was that of England and Wales. Its new structure was introduced on March 31, 1990.[12]

[7] See Wilson (2002), pp. 1321–1325. Chao and Peck (1996) present a market model incorporating electricity losses and transmission congestion.

[8] See Wilson (2002), pp. 1325–1327. Twenty percent of electricity in mature markets is traded day-ahead, less than 10% spot, and the rest long-term. See Wilson (2002), p. 1326.

[9] Singh (1999) discusses e.g. auctions for reserves and other services in an electricity market.

[10] See Joskow (1997), p. 119.

[11] More about the market restructuring can be found in Spiller and Martorell (1996), pp. 106–119.

[12] See Beharrell (1991), p. 45.

The discussion in the literature concerning the electricity market in England and Wales often neglects some features. A short overview is therefore given here. One institution, the Central Electricity Generating Board (CEGB), owned the high-voltage transmission system and most of the power plants before the reform in 1990. It had a monopoly over the wholesale market. Twelve Area Boards with a monopoly for their regions controlled the retail market.[13]

The reform split the CEGB, passing the ownership of power plants to new companies. Nuclear Electric owns all nuclear power plants with a capacity share of about 15%. In 1991 privatization resulted in two new companies, National Power and PowerGen. National Power had a market share of about 52% and PowerGen 33%. The newly established National Grid Company PLC (NGC) owns and manages the power grid. The Area Boards were renamed Regional Electricity Companies (RECs) and privatized in December 1990. They are responsible for the local electricity grid and jointly own the NGC, the system operator.[14]

The sale of electricity to consumers was staged. Customers with a peak demand below 10 MW bought from local distribution companies at regulated tariffs. They received the freedom to choose another power provider depending on their peak demand. In 1998 the retail market was open to full competition. Customers with a peak demand over 10 MW had to pay the competitive prices of the newly created power pool.[15]

The spot market for wholesale electricity, called the power pool, was the key element of the reform. It was designed as a market with 48 separate auctions for each half hour of the next day. Producers had to notify the NGC of their available generation sets and the required prices for each of them. A price bid applied for all 48 auctions. The withdrawal of generation sets was always possible. Information about demand came from RECs and large consumers. Only the latter were allowed to specify a maximum price per half hour. RECs were presumed to be price takers. Based on this information, the NGC prepared a forecast of electricity demand for every 30 minutes of the next day to fulfill its function as the system operator.[16]

The price which producers received for a delivery time period of 30 minutes was the sum of the system marginal price and a capacity payment. The system marginal price was the uniform price of a one-sided auction, with producers as bidders. The auction was one-sided because demand could not be reduced in accordance with those market rules. The capacity payment was introduced, giving an incentive to install production capacity. Reserves, which are unused production capacities, are necessary for the grid's safety. This payment can therefore be viewed as an option price on reserves and underlines the dual nature of power plants. The capacity payment was calculated as the product of the probability of unsatisfied demand and the difference between the value of lost load minus the system marginal price. The probability depended on the situation in each half hour and could be affected, for example, by a failure of a generation set. The value of unsatisfied demand was set at £ 2 per kWh.

[13] See Vickers and Yarrow (1991), p. 191.

[14] See Vickers and Yarrow (1991), pp. 191–192.

[15] See Vickers and Yarrow (1991), pp. 191–194.

[16] See Beharrell (1991), pp. 55–56.

Consumers paid the electricity price, which producers received, plus a payment for ancillary services. Those payments reflect all costs of the system operator to ensure a balanced electricity flow. The NGC had to be compensated for energy losses and, for example, higher costs due to transmission congestion.[17]

A significant part of the literature refers to the electricity market in England and Wales. The attractiveness of the market may arise from the implemented auction system and the effective duopoly of producers. Although three big producers existed, only the privatized companies National Power and PowerGen were influencing the prices. The reason is that Nuclear Electric had the lowest marginal production costs but very high costs to run-up a nuclear generation set. The optimal strategy was to bid zero because demand was always larger than company's production capacity. Hence, another bidder sets the uniform price and Nuclear Electric operated at full capacity.

Wolfram (1998) empirically studies the electricity pool regime in England and Wales. Data cover 6 months in the period 1992–1994. The focus is on the bidding behavior of National Power and PowerGen, the two largest producers, because they very often influenced the market prices. Both producers had power plants located at grid nodes, which were often affected by congestion on the power grid. This made it necessary to call those power plants into action and gave bidders an incentive to submit high bids for these production facilities.[18] An important measure of the study is the mark-up calculated as bid minus marginal costs. Empirical results suggest that the largest producer National Power will bid higher than PowerGen for similar power plants.[19] It shows that the largest company can exercise market power.

Rising prices and investigations showing that producers exercised market power led to a redesign of the electricity market in 2001. Since then, bidders sign bilateral agreements about quantities and prices. The market is therefore similar to a discriminatory auction.[20] Bower and Bunn (2001) study both market designs, before and after 2001, using agent-based computer simulations. The agents of the simulation represent producers of the real electricity market with portfolios of power plants. They have information only about their private portfolio and use internal decision rules to decide about bids for each power plant. Their objectives are profit maximization and a target utilization of the portfolio for each simulated trading day.[21] Consumers are represented by an aggregated demand curve. Demand varies from hour to hour based on real data. Uncertainty is incorporated by adding a random amount of electricity to the basic demand.[22]

[17] See Vickers and Yarrow (1991), pp. 195–196 and Beharrell (1991), pp. 56–59. Note that the formula given by Beharrell (1991), p. 57, is wrong. Newbery and Green (1996) present the correct formula on p. 62. They also give a good overview of the development of the English electricity industry starting in the 1850s.

[18] See Wolfram (1998), p. 716.

[19] See Wolfram (1998), p. 719.

[20] Thomas (2006) gives an overview of the market development since deregulation in 1990.

[21] See Thomas (2006), pp. 573–575.

[22] See Thomas (2006), pp. 575–576.

Bower and Bunn (2001) analyze a uniform-price auction and a discriminatory auction. Agents submit 24 separate bids for each hour of a day in one setting and one bid for the whole day in another setting. It is presumed that the whole available production capacity of a power plant is offered. The agents are only informed about the success of their own bids.[23] The simulation contains 750 days. Only the final 250 days are used for summary statistics. Peak prices of uniform-price auctions are lower than peak prices of the corresponding discriminatory auction. This relationship remains generally unchanged when comparing the prices for all hours. Power plants with low production costs face a higher probability of being underbid by power plants with higher costs in discriminatory auctions than in uniform-price auctions. That is, resources are less efficiently used in discriminatory auctions. Furthermore, the results of the simulations show that submitting one bid for the whole day induces lower prices compared with bidding separately for each hour of a day.[24]

2.3 The Supply-Function Approach

Klemperer and Meyer (1989) developed an approach, which is often used for analyzing electricity markets. They presume an oligopoly of producers facing uncertain but price-elastic demand.Individual marginal costs increase with the quantity of produced goods. Cost functions are common knowledge. Each producer chooses a supply function, which specifies for all possible prices how much he is willing to sell. All producers do this simultaneously before demand is realized. After the realization, the quantity each producers sells is the result of his supply function evaluated at the uniform price. This price is the market price where supply matches demand.[25]

Authors following this approach restrict their analysis to uniform-price auctions. More importantly, they presume common knowledge about production costs. This assumption is questionable, especially for a competitive electricity market. A lot of this information, such as fuel prices and technical conditions influencing the efficiency of a power plant, is actually private. Auctions do not use private information directly, although it may play a roll in calculating bids. But this is done secretly by each producer bidding in the auction. A main concern for them is how to handle incomplete information about competitors. Hence, the supply-function approach is not an appropriate approach for analyzing electricity auctions because it neglects an important issue of a competitive electricity market.

Bolle (1992) refers to the electricity market in England and Wales. He attempts to assess whether a spot market can create complete competition. The theoretical analysis neglects all technical issues of a power market. The author presents three games, which should clear a market of uncertain aggregated demand. Producers are presumed to have constant marginal costs and offer supply functions to the system

[23] See Thomas (2006), pp. 572–573.

[24] See Thomas (2006), pp. 577–582.

[25] See Klemperer and Meyer (1989), pp. 1245 and 1250.

operator who is responsible for the market. Such a supply function assigns to each price a quantity that the bidder is willing to produce. Demand is presumed to fall linearly with higher prices. This assumption may not reflect the market in England and Wales very well because the fraction of large consumers, who were solely able to submit price-dependent bids, was small compared to RECs.

In the first game proposed by Bolle (1992), the system operator announces a constant sale price for consumers before producers offer their supply functions. He always has to operate at zero expected profits. This game does not have an equilibrium. High profits for producers are possible, which contradicts balanced payments of the system operator. The next game reverses the steps of the first game. One arguable assumption is that now the system operator sets the price for consumers before demand is known but settles all transactions using the realized demand. This game has equilibria. Unfortunately, the supply functions decrease with the price. This is an economic result contrary to intuition.

The last game is based on the supply-function approach of Klemperer and Meyer (1989). The game reflects a spot market better because now consumers and producers face the same price after demand is realized. Results show that the spot price is the lowest for the lowest possible demand and the highest – which is the monopoly price here – for the highest demand. Bolle (1992) concludes that a pure spot market may fail to enforce complete competition. It is in turn an argument for the market design in England and Wales because this was not a pure spot market.

Another work analyzing the electricity market in the U.K. is Green and Newbery (1992). They use the supply-function approach and presume a duopoly reflecting the competitive situation at the time of their analysis. The spot market is modeled ignoring capacity payments. Behavioral incentives from repeated auctions are excluded by focusing on a one-shot game.[26] Supply functions are derived. The first result refers to producers with no capacity constraints. The other result is more relevant for the analyzed electricity market because it takes into account that one of the two producers has a capacity constraint. It reflects the situation that National Power had much more installed production capacity than PowerGen.[27]

In a next step, Green and Newbery (1992) fit their theoretical model to real market data. They compare their derived solution with a linear demand curve and marginal costs using data from 1988–89. It highlights the difference between the modeled spot market duopoly and complete competition. Depending on assumed figures, the loss to society is calculated as 6% of the total market revenue under the marginal-cost pricing regime. It is not surprising that calculated deadweight losses increase when the price elasticity of consumers decreases.[28] The case of entrants is additionally reviewed. Green and Newbery (1992) conclude finally that the government underestimated the exercise of market power. A more important result is that five companies with equal production capacities would have led to a much more competitive market compared to the established one with two unequal companies.

---

[26] See Green and Newbery (1992), pp. 932–934.

[27] See Green and Newbery (1992), pp. 937–940. It may be that the solution can only be found by numerical integration. See there on p. 940.

[28] See Green and Newbery (1992), pp. 941–946.

Baldick and Hogan (2002) review at the electricity market in England and Wales in 1999. They use the supply-function approach with linear demand and marginal costs similar to the empirical part of Green and Newbery (1992). In contrast to other works, constrained production capacities, price caps and reserve prices are presumed.[29] The first part of the work discusses problems finding non-decreasing supply functions by solving differential equations as proposed in Klemperer and Meyer (1989), Green and Newbery (1992), and Green (1996). Baldick and Hogan (2002) focus hereby on cases of more than two producers, constrained production capacity and asymmetric costs.[30] Multiple equilibria are problematic because only one result is empirical observed. The stability of equilibria is discussed and conditions for unstable supply-function equilibria are presented.[31]

The constraint of non-decreasing supply functions is important in practice. Hence, it is discussed by Baldick and Hogan (2002) in more detail. They show that under certain conditions the non-decreasing constraint must be explicitly modeled.[32] An iterative numerical approach and its results for different markets conditions are finally presented. The range of stable supply-function equilibria is very small under tight conditions and a binding price cap.[33]

Submitting linear or piecewise linear supply functions is analyzed by Baldick, Grant, and Kahn (2004).[34] Similar to Green and Newbery (1992), linear demand and marginal costs are presumed.[35] In the case of unrestricted production capacities, producers reveal the true intercepts of their marginal costs.[36] The case of constrained capacities is also analyzed. Unfortunately, only producers who are price-takers are presumed to have such constraints and bid their marginal costs. Hence, all producers must know in advance the final market price to anticipate the residual demand, which can be divided among the unconstrained producers. Nevertheless, Baldick, Grant, and Kahn (2004) develop a piecewise linear supply function that is non-decreasing. They emphasize that their result cannot represent an equilibrium among nonlinear functions. Additionally, they presume that demand is unlikely to be realized within a certain small range. Despite these problems, the authors explain reasons why the mentioned difficulties are not important for practical use.[37]

They use their model to compare the market situation in England and Wales from 1996 to 1999 when the duopoly changed to an oligopoly with three or five firms. The calculations show that prices declined as more competitors entered the

---

[29] See Baldick and Hogan (2002), pp. 6–9.

[30] See Baldick and Hogan (2002), pp. 11–28.

[31] See Baldick and Hogan (2002), pp. 29–40.

[32] See Baldick and Hogan (2002), pp. 40–56.

[33] See Baldick and Hogan (2002), pp. 56–110.

[34] Baldick, Grant, and Kahn (2004) assume a strict linear supply function at p. 149. This assumption is relaxed later, see e.g. p. 152.

[35] See Baldick, Grant, and Kahn (2004), p. 148.

[36] See Baldick, Grant, and Kahn (2004), pp. 149–151. Rudkevich (1999) developed a similar model but with zero intercept or marginal costs, see Rudkevich (1999), p. 6.

[37] See Baldick, Grant, and Kahn (2004), pp. 154–156.

market. The predicted prices using the model with positive intercepts of marginal costs are closer to realized market prices than models with zero intercepts such as Green (1999) or Rudkevich (1999).[38]

2.4 Literature Based on von der Fehr/Harbord (1993)

A second approach is based on the work of von der Fehr and Harbord (1993). They presume that a bid consists of a price for an amount of electricity, which is positive and not infinitesimal. The latter is implicitly assumed by the supply-function approach due to modeling differentiable supply functions. This contradicts the bidding rules of the uniform-price auction in the electricity market of England and Wales. A bid in von der Fehr and Harbord (1993) has just one price for a positive quantity. That is, the main difference of both approaches lies in the assumption about the offered amount of electricity.[39]

The focus of von der Fehr and Harbord (1993) is the electricity market of England and Wales a short time after deregulation. Hence, they analyze the bidding behavior of two competitors. Both have constant marginal costs and a production capacity, which cannot be exceeded. The costs and the production capacities are common knowledge. Demand is aggregated from all consumers. It is a random variable with a known lower and upper demand limit. The auction starts with the simultaneous submission of bids for the whole production capacity of each producer. Demand is realized thereafter. Now the auctioneer calculates the lowest price at which supply matches demand. All successful producers receive the same price paid. The authors analyze theoretically the duopoly market by searching for Nash equilibria.[40]

Each producer owns just one power plant with known production capacity. Marginal costs are presumed constant. Demand is uncertain and varies within known ranges. A first result is that pure equilibria exist if the number of bidders needed to match demand is known before the auction.If demand allows only one producer to produce, the paid price is the costs of the competitor and the equilibrium is unique. The highest acceptable price for the auctioneer, the reserve price, is paid in cases where both competitors are needed. Now a continuum of pure equilibria exists because producers only have to bid not higher than the reserve price.[41] Situations, where the number of selected bidders is not certain before the auction takes place, lead to a unique but mixed-strategy Nash equilibrium.[42] The last part of the paper takes a look at real data from July 1990 to April 1991. The authors find changes of bidding behavior over time, which can be seen to conform to the theory. Another

[38] See Baldick, Grant, and Kahn (2004), pp. 159–162.

[39] See von der Fehr and Harbord (1993), p. 532.

[40] See von der Fehr and Harbord (1993), pp. 532–533.

[41] See von der Fehr and Harbord (1993), pp. 533–535.

[42] See von der Fehr and Harbord (1993), p. 536.

explanation takes into account the existence of contracts of differences. They were signed before the deregulation and expired over time.[43]

Brunekreeft (2001) extends the theoretical model of von der Fehr and Harbord (1993). He allows more than two producers owning more than one power plant. Marginal costs are constant. The power plant with the next lowest or highest costs is owned by competitor.[44] Because demand is presumed uncertain, only lower and upper limits for bids are derived.[45] An equilibrium with pure strategies does not exist. It is a similar result to that of von der Fehr and Harbord (1993).[46] The other part of the article deals with real data of the electricity market in England and Wales. The results show that the derived bidding behavior fairly well matches the data provided by von der Fehr and Harbord (1993). Another interesting aspect is that fewer bidders induce higher bids also when the aggregated production capacity remains constant.[47]

Crawford, Crespo, and Tauchen (2007) extend the assumptions of von der Fehr and Harbord (1993). Their goal is the analysis of bidding behavior in a duopoly where producers own many power plants with different costs. All information such as marginal costs is common knowledge. Each producer has the same number of power plants with the capacity of one electricity unit. Marginal costs follow a step function with increasing values. Producers submit bid functions representing the price asked for each power plant.[48] In the analysis, producers are divided either into a price-setter or a non-price-setter. The first is the marginal producer who determines the market price with his bid. The bid of a non-price-setter is lower than the bid of the marginal producer. Hence, it does not affect the market price but allows the non-price-setter to sell electricity. The analysis can only offer bounds for those parts of the bid function which represent utilized power plants not being used for the price calculation. Pure Nash equilibria exist but they do not have to be unique.[49] The theoretical results are used to analyze real data of the electricity market of England and Wales from January 1993 to December 1995. Price-setters show much lower mark-ups than non-price-setters.[50]

The work of Fabra, von der Fehr, and Harbord (2002) is motivated by the discussion of changes in the market structure of England and Wales and problems in the Californian electricity market. The debate refers to the choice of pricing regime similar to the discussion about Treasury bill auctions in the United States.[51] The authors analyze an electricity market with a uniform-price, a discriminatory, and a generalized second-price auction. The last auction type is derived from the idea of

[43] See von der Fehr and Harbord (1993), pp. 539–544.

[44] See Brunekreeft (2001), pp. 104–106.

[45] See Brunekreeft (2001), pp. 107–108.

[46] See Brunekreeft (2001), pp. 108–109.

[47] See Brunekreeft (2001), pp. 109–111.

[48] See Crawford, Crespo, and Tauchen (2007), p. 1239.

[49] See Crawford, Crespo, and Tauchen (2007), pp. 1241–1242.

[50] See Crawford, Crespo, and Tauchen (2007), p. 1255.

[51] The debate on the Treasury Bill auctions started in the 1960s. Institutional aspects are discussed e.g. by Brimmer (1962), Goldstein (1962), and Friedman (1963).

the Vickrey auction.[52] The calculation of the payment for successful bidders forces the producers to bid their true marginal costs in equilibrium. Revealing true costs may be of interest to consumers. The analysis of all three auctions is also done for demand, which is known before calculating bids, as well as for uncertain demand which is realized after submitting bids. All other assumptions follow von der Fehr and Harbord (1993). A similar work to Fabra, von der Fehr, and Harbord (2002) is Fabra, von der Fehr, and Harbord (2006). It comes to similar results but does not analyze the generalized second-price auction.

Results for a market where demand is known before the auction takes places are similar to von der Fehr and Harbord (1993). The uniform-price and discriminatory auction have the same optimal bidding strategy when one producer can fully serve demand. The producer with the lowest costs bids the marginal costs of his competitor. The results for the uniform-price auction are the same as in von der Fehr and Harbord (1993). In the discriminatory auction, no equilibria with pure strategies exist for all other demands. Only unique equilibria with mixed strategies can be derived.[53] Producers in the generalized second-price auction bid their marginal costs independently, whether demand is known before or after the auction. The uniform-price and the discriminatory auction result in equilibria with mixed strategies only for uncertain demand.[54]

Fabra, von der Fehr, and Harbord (2002) and Fabra, von der Fehr, and Harbord (2006) also present bidding strategies as well as a welfare analysis, especially for certain demand. The generalized second-price auction always leads to lowest production costs because the most efficient producer is chosen first to produce electricity. This is reached in the uniform-price or discriminatory auction only if one producer can serve demand. In all other cases, it is possible that either the uniform-price auction or the discriminatory auction results in lower aggregated production costs.[55] The generalized second-price auction weakly dominates the other auction type in terms of consumer surplus if the producer with the lowest marginal costs also has the lowest production capacity. In this case, the uniform-price auction does not lead to better results than the discriminatory auction. The ranking of the auctions varies in the case that the producer with the lowest marginal costs has the same or a higher production capacity compared to his competitor. Only the discriminatory auction leads to an equal or higher consumer surplus compared with the uniform-price auction.[56]

One extension of both works is when bidders do not stick to one bid price for the whole production capacity in the case of certain demand. They are allowed to submit

---

[52] The Vickrey auction refers to the auction introduced by Vickrey (1961).

[53] See Fabra, von der Fehr, and Harbord (2002), pp. 14–18 and Fabra, von der Fehr, and Harbord (2006), pp. 26–27.

[54] See Fabra, von der Fehr, and Harbord (2002), pp. 24–26 and Fabra, von der Fehr, and Harbord (2006), pp. 34–35.

[55] See Fabra, von der Fehr, and Harbord (2002), pp. 18–20 and Fabra, von der Fehr, and Harbord (2006), pp. 27–28.

[56] See Fabra, von der Fehr, and Harbord (2002), pp. 20–21 and Fabra, von der Fehr, and Harbord (2006), pp. 27–28.

step offer-price functions. The existence of a unique equilibrium for the generalized second-price auction is not surprising. Equilibrium outcomes of the other auctions are independent of the step size of the offered supply functions.[57]

Crampes and Creti (2003) analyze a uniform-price auction with similar assumptions as von der Fehr and Harbord (1993). The authors presume additionally that the producer with the lowest marginal costs has the highest production capacity. The analysis focuses on the decision concerning the availability of production capacity if demand is common knowledge. Producers first decide the amount of electricity they want to sell. The bid price of electricity is chosen thereafter. This is done knowing the decisions of all competitors about their available production capacities.[58]

Crampes and Creti (2003) firstly analyze the competition in the second stage. The results show, similar to von der Fehr and Harbord (1993) and Fabra, von der Fehr, and Harbord (2002), that multiple equilibria are possible.[59] This makes the analysis of the capacity competition at the first stage difficult. Multiple equilibria exist if demand can be served by the producer with the highest capacity. A capacity withholding of at least one producer occurs if demand can be served by the producer with the lowest capacity. Otherwise, a withholding is possible but not always sure. In the case that both producers are necessary to serve demand, a capacity withholding is likely. Demand will always be matched, i.e. possible withholding does not lead to a shortage of electricity.[60]

2.5 Related Literature on Auctions

Various approaches are used in the literature concerning electricity markets, of which auction theory is just one. The advantage of this approach is that the market microstructure of auction systems can be modeled very precisely. The focus of this work is on power exchanges, which use auction systems. Hence, auction theory is a good choice for theoretical analysis.

Many different auctions are discussed in the literature. Based on the properties of electricity discussed in Section 2.1, an electricity auction can be characterized as a multi-unit auction. The traded good is homogeneous and divisible into infinite objects of infinitesimal size. Auction rules may define electricity as multiple objects of finite size. Electricity auctions can also be interpreted as multi-unit auctions with complementarities.

Wilson (1979) studies an auction where fractions of one good are sold. Buyers submit bid functions representing the price for all possible shares of the good. The price for all successful bidders is the same. The result of the common value model

[57] See Fabra, von der Fehr, and Harbord (2002), pp. 22–24 and Fabra, von der Fehr, and Harbord (2006), pp. 30–31.

[58] See Crampes and Creti (2003), pp. 5–11.

[59] See Crampes and Creti (2003), pp. 12–17.

[60] See Crampes and Creti (2003), pp. 17–25.

is that an auctioneer's revenue is much lower in this auction type than in a standard first price auction of the whole good.[61]

The assumptions of Ausubel and Cramton (2002) are similar to Wilson (1979). The main difference is that bidders have private values for all fractions of the good.[62] The authors find that a uniform-price auction does not always lead to efficient equilibria, where bidders with the highest valuations get their optimal amounts, because demand is reduced due to bid shading.[63] On the other hand, Ausubel and Cramton (2002) show that an efficient equilibrium can be achieved with a discriminatory auction in cases where the uniform-price auction fails. But it is also possible that this auction does not have an efficient equilibrium.[64] According to Ausubel and Cramton (2002), a ranking of both auction types is ambiguous in terms of efficiency as well as in terms of revenue.[65]

Bikhchandani (1999) analyzes first price auctions where several heterogeneous objects are sold simultaneously. Bidders have reservation values for all combinations, which increase with the number of bought objects. Bidders are not financially constrained.[66] Bikhchandani (1999) shows that a pure strategy Nash equilibrium exists if, and only if, a Walrasian equilibrium exists.[67]

Parisio and Bosco (2003) analyze an electricity market containing two types of power plants. The baseload type has lower production costs compared to the peak-load type. A bidder can own one power plant of each type. In contrast to many electricity market models, costs are presumed to be private knowledge. A uniform-price auction within two different environments is analyzed focusing on the competition among peak-load power plants only. In the first environment, one large producer owns both power plants while the others run only a peak-load type. A strong assumption for the bidding behavior of the large producer is that his competitors bid their marginal costs. The optimal bid exceeds marginal costs. Hence, efficient production is most likely not reached. The second environment refers to a duopoly where producers own both types of power plants. Parisio and Bosco (2003) present optimal bid functions for certain as well as for uncertain price-independent demand.

The aim of Elmaghraby (2005) is to find an auction format that always supports efficient production in equilibrium. The cost structure as well as the production capacity is presumed to be common knowledge. Costs are composed of ramp-up and constant variable costs. Ramp-up costs arise only when a power plant is started. Electricity auctioned in the model covers fluctuating demand over a defined time period. Typically, demand is divided into parts with constant amounts and shorter duration than the whole time period. Elmaghraby (2005) calls this auction format vertical. The analysis shows that neither a uniform-price nor a discriminatory auction could always ensure lowest production costs.

[61] See Wilson (1979), pp. 682–684.

[62] See Ausubel and Cramton (2002), p. 7.

[63] See Ausubel and Cramton (2002), p. 11.

[64] See Ausubel and Cramton (2002), p. 18.

[65] See Ausubel and Cramton (2002), p. 19.

[66] See Bikhchandani (1999), pp. 195–196.

[67] See Bikhchandani (1999), p. 203.

A horizontal auction divides the time-dependent demand curve into slots of constant demand with different durations. Each slot is allocated to only one bidder who gets his bid paid. Efficiency is always reached in equilibrium if slots are auctioned according to duration starting with the longest. Additionally, bidders must submit bids, which are binding for all slots. Hence, bid revision based on new information is not possible.

Chapter 3
Model

Abstract The modeled electricity market is based on a power exchange with an auction system balancing supply and demand. The auctioneer, who is also responsible for the electricity grid, acts as a proxy for consumers and buys electricity from producers. He can choose among three auction types. The uniform price auction pays every winning generator the same price. In the discriminatory auction, the price of the winner is his own bid. Payments of the generalized second-price auction follow the idea of the second price or Vickrey auction. Generators own one power plant each. Producing more electricity than the production capacity is not feasible. Production costs are private information and affiliated. Affiliated random variables are either independent or positively correlated. The concept of affiliation was introduced to auction theory by Milgrom and Weber (1982). A simple probability density function of affiliated production costs is given. It is used for all examples of the model. The numerical results of the examples are presented graphically.

3.1 Assumptions

The electricity market of the model is based on an auction run by the system operator. The system operator or auctioneer is responsible for the stability and the balance of the electricity network. She[1] acts as a proxy for all consumers and buys electricity from generators. Generators take part in the auction because this is their only opportunity to earn money. Typically, production costs vary between generators. Therefore, generators evaluate electricity privately, which refers to the private value model according to Myerson (1981).[2]

The aim of deregulating electricity markets was to establish competition between generators. Therefore, at least two privately owned generators or utilities

[1] The auctioneer or system operator is referred in the work as "she". This is done to clarify the position of the auctioneer and the bidders. A bidder or generator is hence referred as "he".

[2] An overview on auctions is given by Müller (2001).

S. Schöne, *Auctions in the Electricity Market*, Lecture Notes in Economics and Mathematical Systems 617,

were established in new deregulated markets. The electricity market in England and Wales started with three newly created generators. Only two of them were private. They had production capacities of 29 GW and 18 GW. Nuclear Electric plc., the third company, remained public and ran the nuclear power stations with a capacity of 8 GW. Its production costs were mostly the lowest in the market because the competitors owned coal, gas, and oil power plants with worse cost structures.[3] Hence, only two generators effectively competed with each other.[4]

The model focuses on a duopoly, which adopts this market structure. It represents the simplest competition. This allows an overview on the influence of a limited production capacity.[5]

Assumption 3.1. *Two risk-neutral and symmetric generators exist owning one power plant each.*

A utility is normally owned by many investors and run by a management. Although this can lead to a conflict of interests, it is not the focus of this model. Therefore risk-neutrality is assumed, ignoring individual risk preferences.

Assumption 3.2. *Every generator is interested in maximizing his expected profit as long as he does not suffer an expected loss.*

Maximizing the profit is a well-known motivation in market economies. The government accepted this motive to introduce competition in deregulated electricity markets. Generators are interested in expected values because they are assumed to be risk-neutral.

Assumption 3.3. *A generator is able to produce divisible units of electricity.*

Assumption 3.4. *Each plant has the same production capacity of k electricity units. k is common knowledge and $\in \mathbb{R}^+$. A production higher than k electricity units is not feasible.*

Electricity is by nature a divisible commodity. This allows a generator to divide his production capacity into a part used to produce electricity and a part, which is not needed. Power exchanges trade multiples of a specified amount or tick size of electricity units[6] due to practicable and economic reasons. This model covers such rules. If the production capacity is an exact multiple of the tick size, a small unused fraction will remain. The tick sizes are tiny compared with the production capacity of a single generator. Therefore, this case is not emphasized.

Three different basic types of power plants can be distinguished according to production capacity, cost, and availability. In general, the higher the capacity the

[3] See Green (1991), p. 245.

[4] See Green and Newbery (1992), p. 930.

[5] A market with more than two bidders may also be of interest. Its analysis would be more complicated and therefore, it is left for further research. A model for at least two bidders was, e.g. developed by Kaul (2001). He analyzes the behavior of market makers in a stock market.

[6] The EEX AG running a power exchange in Leipzig uses steps of 0.1 MW. See EEX AG (2007a), p. 5.

lower the variable costs. This economically positive result has a drawback. The larger a plant is the longer it normally takes to bring the power plant back online if it has been shut down. A shut-down can be caused by a failure, the need for maintenance and repair or due to failing demand. Additionally, costs to ramp up a plant are higher for large power plants.

Taking these facts into account, a ranking for various types of power plants has been established. The basic demand for electricity is typically served by large plants, e.g. nuclear power plants, classified as baseload plants. Their aggregated capacity normally counts for less than the normal demand. Combined with the lowest variable costs, this type has such a strong competitive advantage in that it can run almost continuously. This practically negates the large ramp-up costs. Plants of the second type satisfy the remaining demand in normal situations. Electricity demand varies greatly during the course of each day. It is usual for this type of power plant to be occasionally shut down for operational reasons. The last type is necessary in cases of excess demand over the aggregated capacity of the first two plant types.

This ranking is preferred by monopolists who minimize their total costs given the demand of electricity. The introduction of competition, which was a key issue of market deregulation, changed the situation completely. It is so far an open question whether the described order will remain. Nevertheless, competitive advantages based on the cost structure are unchanged. Baseload plants still have the best advantages due to the lowest variable costs. As long as demand exceeds their aggregated production capacity, they operate at full capacity and do not have to deal with high ramp-up costs. The cost advantages of the remaining two plant types suggests that the strongest competition takes place within the group of generators running the same plant type. The model focuses on this situation.[7] Consequently, the same production capacity for the generators can be assumed for reasons of simplicity.

Production capacity is changeable in the long run. Power exchanges, which are the focus of this model, operate auctions for spot markets or forward markets. Both markets have in common that electricity must be delivered immediately or within days. This time horizon is too short for an enlargement of capacities. Additionally, electricity is economically not storable over a long period or in large amounts. Hence, the production capacity k is an important restriction for generators.

Assumption 3.5. *Each generator has constant marginal costs. c_i denotes the costs of generator i to produce one unit. It is the realization of the stochastic information variable C_i, $i = 1,2$.*

Assumption 3.6. *Marginal costs are private information of the generator.*

Competition between generators of the same power plant type is of interest because they are the strongest competitors due to their comparable costs. These generators face similar fixed costs, run-up costs and marginal costs. Fixed costs must be covered by the contribution margin of the entire production. A generator who does

[7] Average marginal costs of power plants running with different fuels for the market in England and Wales are given by Wolfram (1998), p. 711, Fn. 19. The figures show a clear ranking, which supports the assumption.

not expect to earn at least the fixed costs would not participate in the market. Hence, it is no loss of generality to neglect this cost type. Ramp-up costs can be viewed as fixed costs if both generators start with the same status of the power plant, i.e. shut down or running. The model focuses on a one-shot game. Hence, it is reasonable to presume that both generators have shut-down power plants. Marginal costs are the most important part of the production costs of one electricity unit. For simplicity reasons, they are assumed to be constant.

Production costs depend on prices of the input factors as well as on technical factors. Both influences are hardly observable by the public. Varied production costs can be a result of environmental conditions or contract conditions for the purchase of input factors. An obligation to publish such private information would induce incentive problems, which are not of concern here. In the case where private information is announced correctly, the market would become a kind of Bertrand competition and uncertainty about costs would disappear. But just this uncertainty is a key issue of the model.

Assumption 3.7. *All information variables C_i, $i = 1,2$, are affiliated. The probability density function $f(c_j, c_i)$, with $i, j = 1,2$ and $i \neq j$, is twice continuously differentiable and positive. It is symmetric too, i.e.*

$$f(c_i, c_j) = f(c_j, c_i). \tag{3.1}$$

The support is $[\underline{c}, \bar{c}]$, with $\underline{c}, \bar{c} \in \mathbb{R}_0^+$, and $\bar{c} > \underline{c}$. The limits are common knowledge.

Input prices of generators are often positively correlated, e.g. because they use the same fuel. In cases where different fuels are used, such a link between price changes is weaker but still exists. The reason is that fuels can be substituted in this market. A concept that can model such behavior is affiliation given by Milgrom and Weber (1982).[8] It says that the marginal costs of both generators are non-negatively correlated from the position of a generator knowing his own costs. Hence, it also covers the case of independently distributed costs.

Whereas the information about individual costs is private, the type of power plant is common knowledge. This is sufficient to predict the range of marginal costs. Only one cost range exists in the model because competition between generators of the same type is analyzed. The upper boundary is represented by $\bar{c}$ and the lower by $\underline{c}$. $\bar{c} > \underline{c}$ allows competition. Otherwise, Assumption 3.6 is useless.

Assumption 3.8. *Generators do not face any transaction cost.*

Assumption 3.9. *Aggregated demand for electricity $d \in \mathbb{R}^+$ is price independent and common knowledge.*

Deregulating the electricity market can be accomplished in various way, especially when viewed from the consumers' point of view. One concept is to allow consumers to buy power from every generator directly. For economic reasons, the market would probably be split up. Some consumers would prefer offers with fixed

[8] The definition of affiliation is given by Milgrom and Weber (1982), p. 1118.

prices for a unit of electricity. Other consumers may negotiate contracts individually, stipulating preferred price, quantities or supply dates. A market with such a structure can be found in Germany. But this market structure is not focused on here. Therefore, the model does not cover this particular approach.

Nevertheless, markets with bilateral trading may have power exchanges based on auctions. Such institutions are often used as a reference for individual contracts. Despite their potential small fraction of traded electricity, power exchanges can have an important influence on prices and allocations. The model may give an interesting starting point for the analysis of this markets.

Quite a different approach was used in the electricity market in England and Wales. The electricity market was separated in a wholesale market and a retail market. All generators producing electricity had to sell in the wholesale market. It was designed as a one-sided auction. The auctioneer acted for the local suppliers buying the aggregated demand. The local companies resold the electricity to the consumers.[9]

An interesting feature of this market structure is that essential technical conditions could be handled with economic methods. Electricity supply and demand has to be balanced in the grid continuously. It is impossible to separate individual power flows. Hence, only aggregated figures are relevant. Auctions fit these technical needs very well, which is a good reason to follow that market approach in the model.

Electricity is a vital commodity for private households and companies. Contracts for companies can contain fluctuating electricity prices. It allows companies to act on actual prices. Such options are unusual for households.[10] Due to the importance of electricity, it is assumed that having it is more important than the price paid. This results in price-independent demand.

Generators have always attempted to forecast the demand of electricity. Such information is crucial for scheduling the usage of each power plant a company owns. It is a vital tool for keeping down costs. The system operator also needs this information to ensure the safety of the electricity grid. The long-term experience and technical improvements allow a very precise forecast of short-term demand. Hence, it seems acceptable to abstract from demand uncertainty.[11]

Assumption 3.10. *Supply and demand are located at the same node in the electricity grid.*[12]

Balancing supply and demand in an electricity grid is a difficult task. Electricity has the property of transforming into heat as it flows from the producer to the consumer. Another problem can result from the geographical allocation of supply

---

[9] The market structure after deregulation in England and Wales is explained e.g. in Green (1991).

[10] A test with private households in Germany showed the possibility of implementing real-time electricity prices for this group of consumers. The basis was a real-time information system that allowed price dependent behavior. An implementation was not recommended because costs outweighed savings. See Möhring-Hüser, Morovic, and Pilhar (1998).

[11] A justification for the assumption of certain demand can be found e.g. in Borenstein (2002).

[12] Another alternative with the same result is to assume that transmission losses do not exist and that the power flow is not restricted, e.g. by transmission line capacities. Supatgiat, Zhang, and Birge (2001) use this approach, see there on p. 98.

and demand in the grid. The transmission system has capacity limits. If they are reached, stability can only be achieved by separating the market into local markets. This allows attracting more demand on one side and more supply on the other side of the congestion until balance is reached.[13] The model should deal with economic issues. Assumption 3.10 results in abstracting from described technical issues.[14] Consequently, aggregated demand has to be matched with supply only.

Assumption 3.11. *The auction is the only place to trade electricity. One auction takes place.*

A spot market, sometimes called a day-ahead market, usually runs several similar auctions with the same duration but with different delivery times on the next day. Time slots of one hour or 30 minutes are normally auctioned. Auctions are cleared simultaneously. Electricity is not storable. Hence, each auction starts with the same conditions except the state of the power plant (shutdown or running). Ramp-up costs, which can cause serious decision problems for bidders, are not of concern. Therefore, it is no loss of generality to focus on a single auction.

Assumption 3.12. *The auctioneer sets a reserve price $\bar{p}$, with $\bar{p} \in \mathbb{R}^+$ and $\bar{p} > \bar{c}$. She announces $\bar{p}$, the highest price she is willing to pay for one electricity unit in advance of the auction start.*

The auctioneer considers the needs of consumers. Because demand is assumed to be price-independent, generators may require prices which are too high. The price cap $\bar{p}$ respects this need. Consumers' reserve price $\bar{p}$ lies above the highest possible marginal costs. This also gives the most inefficient generator a chance to earn money by taking part in the auction and therefore to stay in the market.

Because the auction takes place only once, due to Assumption 3.11, and because $\bar{p}$ is common knowledge, generators are deterred from bidding higher than the price cap. Consequently, all bidders are eager to offer bids, which are acceptable to the auctioneer. This ensures the highest offered production capacity. Hence, it prevents electricity shortages for consumers due to strategic behavior by generators.

Assumption 3.13. *The auctioneer announces the auction type, which is used in the auction, before the submission period for bids begins. The set of auction types contains the discriminatory auction, the uniform-price auction, and the generalized second-price auction, a variant of the second price auction.*

The rules of the auction types mentioned above are described in Section 3.2, see Definitions 3.1–3.3.

Assumption 3.14. *Bidding is done in a non-cooperative manner.*

[13] Bohn, Caramanis, and Schweppe (1984) develop a model that incorporates transmission losses and limited transmission capacity. They refer – in contradiction to this model – to a market with a welfare-maximizing monopolist.

[14] Neame, Philpott, and Pritchard (1999) use this assumption. See there on p. 2.

Assumption 3.15. *All aspects of the market and the auctions are common knowledge.*

Assumption 3.16. *Bidding generators take the behavior of their competitors into account and act strategically.*

A generator possesses information about the marginal costs of his competitor. The information is incomplete, but ignoring it means wasting some advantage. Such behavior would lead to lower revenue on average, which contradicts Assumption 3.2. Hence, the auction can be analyzed using the concept of the Bayes–Nash equilibrium according to Harsanyi (1967/68).[15] Precisely, the symmetric equilibrium (b^*, b^*) is searched for, where b^* denotes the optimal bidding strategy of a generator. b^* maximizes generator's expected profit assuming competitors' bidding strategies are given and correct.

3.2 Definitions

Before an auction takes place, the auctioneer announces the auction type. Hence, every generator knows the auction rules before he decides on his bid. Three different auction types are analyzed, the definitions of which are given for the two-bidder case as follows.

Definition 3.1. *The discriminatory auction is described by the following set of rules.*

1. *A bidder submits a bid for delivery of one electricity unit as sealed bid.*
2. *Bidders offer their whole production capacity.*
3. *Only non-negative bids smaller than $\bar{p}$ are accepted.*
4. *The bidder with the lowest bid sells electricity until his production capacity is fully utilized. The second bidder is successful with his bid only in the case of remaining demand. Draws are solved by rationing pro rata of offered supply.*[16]
5. *A successful bidder gets his bid paid.*

A generator bidding under the regime of this auction type can calculate his profit in advance for the case of winning or losing. Because his own bid determines the profit directly, he has to find a balance between a high price and a good chance of selling the most electricity.

Definition 3.2. *The rules of the uniform-price auction are the same as those of the discriminatory auction except for the prices paid. Successful bidders are paid a uniform price. It is determined by the highest bid of all successful bidders.*

The advantage of this auction type is that no bidder is discriminated against. The price paid per unit is the same for every scheduled generator. This characteristic

[15] Good explanations of the Nash equilibrium and the Bayes–Nash equilibrium are given in Fudenberg and Tirole (1995), pp. 11–42 and pp. 215–226 respectively.

[16] Another opportunity for handling equal bids is to use a lottery. The probability of being selected is the same for everyone.

is important in discussions about the fairness of a pricing system. The final auction price can be viewed as the best available market price for electricity because it matches supply and demand at the lowest possible price.

Definition 3.3. *The rules of the generalized second-price auction are the same as those of the discriminatory auction except for the prices paid. They depend on demand as follows.*

1. *Suppose demand is satisfied by the bidder with the lowest bid. In this case, he is paid the bid of the unsuccessful bidder.*
2. *Suppose demand is sufficient such that both bidders fully utilize their production capacities. They are paid the reservation price.*
3. *In all other cases, the bidder with the highest bid gets the reservation price paid. The bidder with the lowest bid receives the reservation price for the amount of electricity that his competitor sells. The price for the remaining electricity is the bid of the competitor.*

This auction adopts the idea of the second price auction or Vickrey auction. This auction type forces truth telling bids but is rarely used.[17] This makes it an interesting candidate for analysis. The basic principle of the auction is to pay the winning bidder the price the auctioneer would have had paid if the winner had not taken part. Competitors, who had been replaced the winner, would have been paid their own bid for unused production capacity.[18] The advantage from the bidders' point of view is that the offered bid is not directly used for setting the payments. Hence, bidder's most important concern is to maximize the probability of winning.

A general characteristic of auctions is that the auctioneer selects the bidders in an order, which minimizes her costs given the offered bids. This matches the objective of the auctioneer as a proxy for consumers, who want to pay the lowest price.

3.3 Preliminaries

Assumptions about the underlying distribution function are important for the analysis. The following definitions and mathematical terms are given for clarification and to simplify the work.

Definition 3.4.

(a) *The marginal distribution is given by*

$$f(c) \quad \equiv \int_{\underline{c}}^{\bar{c}} f(\tau, c)\, d\tau. \tag{3.2}$$

[17] See Rothkopf, Teisberg, and Kahn (1990), pp. 94–95.

[18] Fabra, von der Fehr, and Harbord (2002) use e.g. this rule for the generalized second-price auction. See there on p. 12.

(b) *The cumulative marginal distribution function is defined as*

$$F(c) \equiv \int_{\underline{c}}^{c} f(\tau)\, d\tau, \qquad \text{with} \quad F(\bar{c}) = 1. \tag{3.3}$$

(c) *The conditional probability that generator j has marginal costs of c_j given generator i having marginal costs of c_i, with $i \neq j$, is obtained by*

$$f(c_j|c_i) \equiv \frac{f(c_j, c_i)}{\int_{\underline{c}}^{\bar{c}} f(\tau, c_i)\, d\tau} = \frac{f(c_j, c_i)}{f(c_i)}. \tag{3.4}$$

(d) *The conditional distribution function is determined by*

$$F(c_j|c_i) \equiv \int_{\underline{c}}^{c_j} f(\tau|c_i)\, d\tau = \frac{\int_{\underline{c}}^{c_j} f(\tau, c_i)\, d\tau}{f(c_i)}. \tag{3.5}$$

(e) *The first partial derivative of the probability density function is*

$$f_i(c_j, c_i) \equiv \frac{\partial f(c_j, c_i)}{\partial c_i}. \tag{3.6}$$

(f) *The first partial derivative of the conditional probability density function with respect to c_i is given by*

$$f_i(c_j|c_i) \equiv \frac{\partial f(c_j|c_i)}{\partial c_i} = \frac{f_i(c_j, c_i) - f(c_j|c_i) \int_{\underline{c}}^{\bar{c}} f_i(\tau, c_i)\, d\tau}{f(c_i)}. \tag{3.7}$$

(g) *The first partial derivative of the conditional distribution function with respect to c_i is given by*

$$F_i(c_j|c_i) \equiv \frac{\partial F(c_j|c_i)}{\partial c_i} = \int_{\underline{c}}^{c_j} f_i(\tau|c_i)\, d\tau = \frac{\int_{\underline{c}}^{c_j} f_i(\tau, c_i)\, d\tau - F(c_j|c_i) \int_{\underline{c}}^{\bar{c}} f_i(\tau, c_i)\, d\tau}{f(c_i)}. \tag{3.8}$$

The following property of an affiliated probability distribution is important for the analysis of the model.

Lemma 3.1.

(a) *The first partial derivative of the conditional distribution function satisfies the following condition*

$$F_i(c_j|c_i) \leq 0. \tag{3.9}$$

(b) *The following inequation holds true in the model*

$$\frac{f_i(c_j|c_i)}{f(c_j|c_i)} + \frac{F_i(c_j|c_i)}{1-F(c_j|c_i)} \leq 0. \tag{3.10}$$

Proof.

(a) According to Assumption 3.7, $f(c_i,c_j)$ is an affiliated probability distribution. That $F(c_j|c_i)$ is nonincreasing in c_i follows directly from the affiliation property. Affiliation refers to first-order stochastic dominance of the conditional distribution function.[19]

(b) The hazard rate is nonincreasing given affiliated distribution.[20] Hence, the following condition holds true

$$\frac{\partial \frac{f(c_j|c_i)}{1-F(c_j|c_i)}}{\partial c_i} \leq 0.$$

The first partial derivative is obtained by

$$\frac{\partial \frac{f(c_j|c_i)}{1-F(c_j|c_i)}}{\partial c_i} = \frac{f_i(c_j|c_i)}{1-F(c_j|c_i)} + \frac{f(c_j|c_i)F_i(c_j|c_i)}{[1-F(c_j|c_i)]^2}.$$

Note the probability density $f(c_j,c_i)$ is positive according to Assumption 3.7. Therefore, the condition is given by

$$0 \geq \frac{f_i(c_j|c_i)}{1-F(c_j|c_i)} - \frac{f(c_j|c_i)F_i(c_j|c_i)}{[1-F(c_j|c_i)]^2}$$

$$\Longleftrightarrow$$

$$\frac{f_i(c_j|c_i)}{f(c_j|c_i)} \leq \frac{(-)F_i(c_j|c_i)}{1-F(c_j|c_i)}.$$

□

An overview of functions concerning an independent probability distribution is given with Lemma 3.2. These expressions represent the basis on which some results can be simplified.

[19] See, e.g. Krishna (2002), p. 271.

[20] See, e.g. Krishna (2002), p. 271.

Lemma 3.2.

(a) *Conditional probability $f(c_j|c_i)$ equals marginal distribution $f(c_j)$ for independently drawn marginal costs*

$$f(c_j|c_i) = f(c_j). \tag{3.11}$$

(b) *The conditional distribution function for an independent probability distribution is determined by*

$$F(c_j|c_i) = F(c_j). \tag{3.12}$$

(c) *The first partial derivative of the conditional probability density function $f_i(c_j|c_i)$ for independently distributed marginal costs is given by*

$$f_i(c_j|c_i) = 0. \tag{3.13}$$

(d) *The first partial derivative of the conditional distribution function $F_i(c_j|c_i)$, given independent distribution, is obtained by*

$$F_i(c_j|c_i) = 0. \tag{3.14}$$

Proof.

(a) For an independent probability distribution holds true

$$f(c_j, c_i) = f(c_j)\, f(c_i)$$

because this results in a conditional probability $f(c_j|c_i)$, which does not depend on c_i. Using (3.4), the conditional probability is given by

$$f(c_j|c_i) = \frac{f(c_j, c_i)}{f(c_i)} = \frac{f(c_j)\, f(c_i)}{f(c_i)} = f(c_j).$$

(b) Using (3.11) and (3.3), conditional distribution function (3.5) for independently distributed marginal costs is rewritten to

$$F(c_j|c_i) = \int_{\underline{c}}^{c_j} f(\tau|c_i)\, d\tau = \int_{\underline{c}}^{c_j} f(\tau)\, d\tau = F(c_j).$$

(c) Taking into account (3.11), the first derivative of the conditional probability density function $f_i(c_j|c_i)$, given independently distributed marginal costs, is obtained by

$$f_i(c_j|c_i) = \frac{\partial f(c_j|c_i)}{\partial c_i} = \frac{\partial f(c_j)}{\partial c_i} = 0.$$

(d) Equation (3.13) is used to derive $F_i(c_j|c_i)$, see (3.8), for independently distributed marginal costs

$$F_i(c_j|c_i) = \int_{\underline{c}}^{c_j} f_i(\tau|c_i)\, d\tau = 0.$$

□

3.4 Example

An example is presented to illustrate theoretical results. More specific assumptions about the probability density function and the reserve price are made to keep results feasible but as general as possible.

Affiliation is a key issue of the model. Its weak form was presumed, which includes independently drawn marginal costs. The strong affiliation refers to strictly positive correlated marginal costs. The distinction is made by setting constant a in Definition 3.5 accordingly.

Definition 3.5. *The following specification of probability density function (3.1) is used in the example*

$$f(c_j, c_i) = \frac{1 + a c_i c_j}{1 + \frac{a}{4}}, \tag{3.15}$$

with a support of marginal costs of

$$c_i, c_j \in [0,1]. \tag{3.16}$$

Constant a determines the degree of affiliation between marginal costs of generators and satisfies following condition

$$a \geq 0. \tag{3.17}$$

The function given with Definition 3.5 must satisfy Assumption 3.7 to be of use as an example. This is shown with Theorem 3.1.

Theorem 3.1. *Suppose the function given by Definition 3.5*

(a) *The function represents a density function of random variables with positive values.*
(b) *The random variables of the function are affiliated.*

Proof.

(a) A density function of random variables must be non-negative at first. This condition always holds true because no factor is negative due to (3.16) and (3.17). A density function must satisfy

$$\int_0^1 \int_0^1 f(c_j, c_i)\, dc_j dc_i = 1.$$

Equation (3.15) is used to get

$$\begin{aligned}\int_0^1 \int_0^1 \frac{1 + ac_i c_j}{1 + \frac{a}{4}}\, dc_j dc_i &= \tfrac{1}{1+\frac{a}{4}} \int_0^1 \left[c_j + \tfrac{ac_i c_j^2}{2} \right] \Big|_0^1 \, dc_i \\ &= \tfrac{1}{1+\frac{a}{4}} \int_0^1 \left[1 + \tfrac{ac_i}{2} \right] dc_i \\ &= \tfrac{1}{1+\frac{a}{4}} \left[c_i + \tfrac{ac_i^2}{4} \right] \Big|_0^1 = 1.\end{aligned}$$

Even the last condition holds true. Hence, (3.15) represents a density function.

(b) Affiliation is given if the following condition holds true[21]

$$\frac{\partial^2}{\partial c_j \partial c_i} \ln(f(c_j, c_i)) \geq 0 \tag{3.18}$$

with

$$\begin{aligned}\frac{\partial^2}{\partial c_j \partial c_i} \ln(f(c_j, c_i)) &= \frac{\partial}{\partial c_i} \frac{f_j(c_j, c_i)}{f(c_j, c_i)} \\ &= \frac{f_{ji}(c_j, c_i)}{f(c_j, c_i)} - \frac{f_j(c_j, c_i) f_i(c_j, c_i)}{f(c_j, c_i)^2}.\end{aligned}$$

The first partial derivative of the density function, see (3.15), with respect to c_j is obtained by

$$f_j(c_j, c_i) = \frac{\partial f(c_j, c_i)}{\partial c_j} = \frac{ac_i}{1 + \frac{a}{4}}. \tag{3.19}$$

The first partial derivative of (3.15) with respect to c_i is given by

$$f_i(c_j, c_i) = \frac{\partial f(c_j, c_i)}{\partial c_i} = \frac{ac_j}{1 + \frac{a}{4}}. \tag{3.20}$$

Using (3.19), the second partial derivative of (3.15) with respect to c_j and c_i is obtained by

$$f_{ji}(c_j, c_i) = \frac{\partial^2 f(c_j, c_i)}{\partial c_j \partial c_i} = \frac{a}{1 + \frac{a}{4}}. \tag{3.21}$$

[21] See Krishna (2002), p. 269.

Because the density function is positive due to Theorem 3.1 (a) and taking into account (3.15), (3.19), (3.20), and (3.21), condition (3.18) is rewritten to

$$
\begin{aligned}
&0 \leq f(c_j,c_i)\, f_{ji}(c_j,c_i) - f_j(c_j,c_i)\, f_i(c_j,c_i) \\
&\Longleftrightarrow \\
&0 \leq \frac{1+ac_ic_j}{1+\frac{a}{4}} \frac{a}{1+\frac{a}{4}} - \frac{ac_i}{1+\frac{a}{4}} \frac{ac_j}{1+\frac{a}{4}} \\
&\;\;\leq \frac{a}{\left[1+\frac{a}{4}\right]^2} \qquad (3.22)\\
&\Longleftrightarrow \\
&0 \leq a.
\end{aligned}
$$

This condition is defined with (3.17). Hence, affiliation is given. □

Remark 3.1. *Probability density function (3.15) with $a = 0$ results in*

$$
f(c_j,c_i) = 1. \qquad (3.23)
$$

Obviously, this represents an independent probability distribution. All other values for a lead to strong affiliation because the right-hand term of the inequation (3.22) for $a > 0$ is always greater than zero.

Important functions concerning the probability distribution of the example are presented here.

Lemma 3.3.

(a) *Marginal distribution (3.2) is obtained by*

$$
f(c_i) = \begin{cases} 1 & : a = 0 \\ \dfrac{1+\frac{a}{2}c_i}{1+\frac{a}{4}} & : a > 0. \end{cases} \qquad (3.24)
$$

(b) *Cumulative marginal distribution (3.3) is obtained by*

$$
F(c_i) = \begin{cases} c_i & : a = 0 \\ \dfrac{1+\frac{a}{4}c_i}{1+\frac{a}{4}}\, c_i & : a > 0. \end{cases} \qquad (3.25)
$$

(c) *Conditional probability density function (3.4) is obtained by*

$$f(c_j|c_i) = \begin{cases} 1 & : a = 0 \\ \dfrac{1 + a c_i c_j}{1 + \frac{a}{2} c_i} & : a > 0. \end{cases} \tag{3.26}$$

(d) *Conditional cumulative distribution function (3.5) is given by*

$$F(c_j|c_i) = \begin{cases} c_j & : a = 0 \\ \dfrac{1 + \frac{a}{2} c_i c_j}{1 + \frac{a}{2} c_i} c_j & : a > 0. \end{cases} \tag{3.27}$$

(e) *The first partial derivative of probability density function (3.15) with respect to c_j is obtained by*

$$f_j(c_j, c_i) = \begin{cases} 0 & : a = 0 \\ \dfrac{a c_i}{1 + \frac{a}{4}} & : a > 0. \end{cases} \tag{3.28}$$

(f) *The first partial derivative of probability density function (3.15) with respect to c_i, see (3.6), is given by*

$$f_i(c_j, c_i) = \begin{cases} 0 & : a = 0 \\ \dfrac{a c_j}{1 + \frac{a}{4}} & : a > 0. \end{cases} \tag{3.29}$$

(g) *The second partial derivative of probability density function (3.15) with respect to c_j and c_i is obtained by*

$$f_{ji}(c_j, c_i) = \begin{cases} 0 & : a = 0 \\ \dfrac{a}{1 + \frac{a}{4}} & : a > 0. \end{cases} \tag{3.30}$$

(h) *Derivative (3.7) of conditional probability density function is determined by*

$$f_i(c_j|c_i) = \begin{cases} 0 & : a = 0 \\ \dfrac{a \left[c_j - \frac{1}{2}\right]}{\left[1 + \frac{a}{2} c_i\right]^2} & : a > 0. \end{cases} \tag{3.31}$$

(i) *Derivative (3.8) of conditional cumulative probability function is given by*

$$F_i(c_j|c_i) = \begin{cases} 0 & : a = 0 \\ (-)\dfrac{a[1-c_j]c_j}{2\left[1+\frac{a}{2}c_i\right]^2} & : a > 0. \end{cases} \tag{3.32}$$

Proof.

(a) Due to symmetry of (3.15), marginal distribution (3.2) is obtained by

$$f(c_i) = \frac{1}{1+\frac{a}{4}} \int_0^1 [1+ac_i\tau]\,d\tau = \frac{1+\frac{a}{2}c_i}{1+\frac{a}{4}}.$$

If $a = 0$ is presumed, the function is simplified to

$$f(c_i) = 1.$$

(b) Taking into account (3.24), cumulative marginal distribution (3.3) is given by

$$F(c_i) = \frac{1}{2+\frac{a}{2}} \int_0^{c_i} [2+a\tau]\,d\tau = \frac{2c_i+\frac{a}{2}c_i^2}{2+\frac{a}{2}} = \frac{1+\frac{a}{4}c_i}{1+\frac{a}{4}}c_i.$$

Given $a = 0$, the function can be simplified to

$$F(c_i) = c_i.$$

(c) Using (3.15) and (3.24), conditional probability density function (3.4) is obtained by

$$f(c_j|c_i) = \frac{\frac{1+ac_ic_j}{1+\frac{a}{4}}}{\frac{1+\frac{a}{2}c_i}{1+\frac{a}{4}}} = \frac{1+ac_ic_j}{1+\frac{a}{2}c_i}.$$

Given $a = 0$, the function is simplified to

$$f(c_j|c_i) = 1.$$

(d) Using (3.26), conditional cumulative distribution function (3.5) is given by

$$F(c_j|c_i) = \int_0^{c_j} \frac{1+ac_i\tau}{1+\frac{a}{2}c_i}\,d\tau$$

$$= \frac{\left[\tau + \frac{a}{2}c_i\tau^2\right]\Big|_0^{c_j}}{1+\frac{a}{2}c_i} = \frac{1+\frac{a}{2}c_ic_j}{1+\frac{a}{2}c_i}c_j.$$

If $a = 0$ is presumed, the function can be simplified to

$$F(c_j|c_i) = c_j.$$

(e) The derivative is still derived with the proof of Theorem 3.1. See (3.19). A simpler result is obtained for $a = 0$

$$f_j(c_j, c_i) = 0.$$

(f) The derivative is still derived with the proof of Theorem 3.1. See (3.20). The result can for $a = 0$ be simplified to

$$f_i(c_j, c_i) = 0.$$

(g) The derivative is still derived with the proof of Theorem 3.1. See (3.21). For $a = 0$, a simpler result is

$$f_{ji}(c_j, c_i) = 0.$$

(h) Derivative (3.7) of conditional probability density function is given by, using (3.26)

$$f_i(c_j|c_i) = \frac{\partial f(c_j|c_i)}{\partial c_i}$$
$$= \frac{ac_j\left[1+\frac{a}{2}c_i\right] - \frac{a}{2}[1+ac_ic_j]}{\left[1+\frac{a}{2}c_i\right]^2} = \frac{a\left[c_j - \frac{1}{2}\right]}{\left[1+\frac{a}{2}c_i\right]^2}.$$

It is easy to see that for $a = 0$ the derivative is reduced to

$$f_i(c_j|c_i) = 0.$$

The general result was presented with (3.13).

(i) Derivative (3.8) of conditional cumulative probability function is given by, using (3.27)

$$F_i(c_j|c_i) = \frac{\partial F(c_j|c_i)}{\partial c_i}$$
$$= a\frac{c_j\left[1+\frac{a}{2}c_i\right] - \left[1+\frac{a}{2}c_ic_j\right]}{2\left[1+\frac{a}{2}c_i\right]^2}c_j = (-)\frac{a[1-c_j]c_j}{2\left[1+\frac{a}{2}c_i\right]^2}.$$

The general result for $a = 0$ is known from (3.14). It is easily checked

$$F_i(c_j|c_i) = 0.$$

□

The conditional cumulative distribution function is important for a generator deciding about his bid in an auction. In all three auctions of this model, the auctioneer selects the bidder with the lowest bid first. Hence, a generator is interested in being chosen first, which occurs in all cases when his competitor has bid higher. The probability of losing the auction is given by $F(c_j|c_i)$ telling how probable it is that the competitor has lower or equal marginal costs. The probability is calculated with all available information, which includes the knowledge about his own cost c_i. Figure 3.1 shows the graph for independently distributed marginal costs. Strong affiliation is presented by Figure 3.2 with $a = 4$.

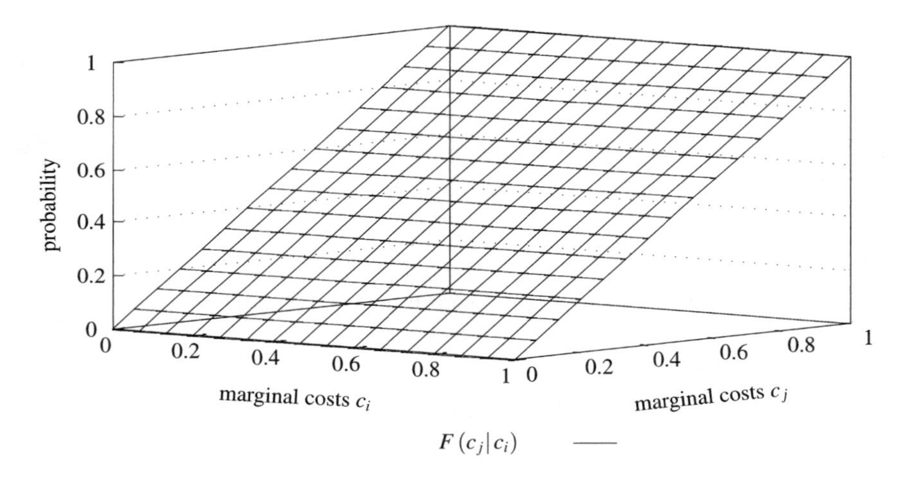

Fig. 3.1 Conditional cumulative distribution for independently distributed costs

It is obvious that Figure 3.1 weakly dominates Figure 3.2. Strong affiliated costs ($a > 0$) first-order stochastically dominate independently distributed costs ($a = 0$) if the following holds true, see (3.27)[22]

$$F(c_j|c_i) \geq F(c_j|c_i) \qquad \forall\, c_i, c_j \in [0,1]$$

$$\Longleftrightarrow \; c_j \geq \frac{1 + \frac{a}{2} c_i c_j}{1 + \frac{a}{2} c_i} c_j$$

$$\Longleftrightarrow \; c_j \leq 1.$$

[22] See Krishna (2002), p. 259. Some other definitions of stochastic dominance are also given there, see pp. 259–263.

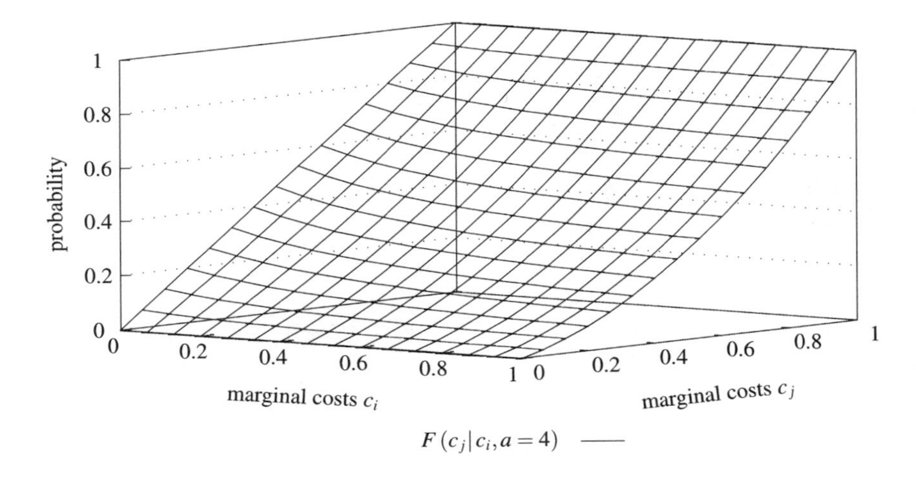

Fig. 3.2 Conditional cumulative distribution for strong affiliation

Obviously, the condition is valid. A derived expression is more relevant for further analysis. It answers the generator's question about the probability of having the lowest marginal costs and knowing his own marginal costs c. The probability is important in cases where the bidder has to balance profits from winning or loosing an auction. The expression is obtained by using (3.5) and (3.27)

$$
\begin{aligned}
E\left[1_{\{c<\tau\}}\right] &= \int_{\underline{c}}^{\bar{c}} f(\tau|c)\,1_{\{c<\tau\}}\,d\tau = \int_{c}^{\bar{c}} f(\tau|c)\,d\tau \\
&= 1 - F(c|c) \\
&= \begin{cases} 1-c & : a=0 \\[2ex] 1-\dfrac{1+\frac{a}{2}c^2}{1+\frac{a}{2}c}\,c & : a>0. \end{cases}
\end{aligned}
\tag{3.33}
$$

The function is shown in Figure 3.3. The line at $a = 0$ represents the graph for independently distributed marginal costs. The values are the same as for the diagonal from point $(0,0)$ to $(1,1)$ in Figure 3.1. The line at $a = 4$ is similarly derived from Figure 3.2. Figure 3.3 demonstrates well, that the probability of winning increases with higher a, i.e. stronger affiliation.

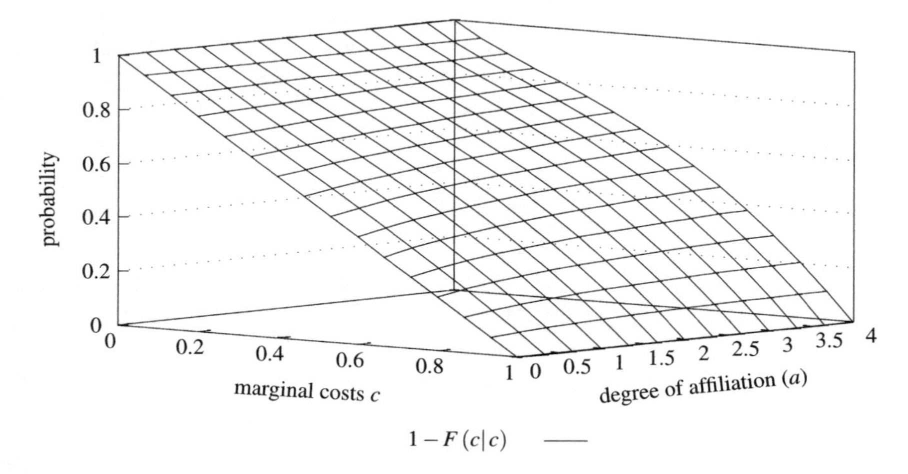

Fig. 3.3 Probability of having lowest marginal costs (generator's view)

The last assumption made for the example is that the auctioneer sets the reserve price as follows

$$\bar{p} = 3. \tag{3.34}$$

A reserve price of 3 is greater than the upper limit of marginal costs $\bar{c}$, which is 1 due to (3.16). Hence, (3.34) satisfies Assumption 3.12.

Chapter 4
Results

Abstract The electricity market modeled in Chapter 3 is the starting point of analyzing the uniform price auction, the discriminatory auction, and the generalized second-price auction. Generators bid differently according to demand. This allows focusing on different demand ranges. The optimal bidding strategy of a generator is derived for each auction type and each demand range. The results are used to determine the expected costs, profits, and contribution margins of a producer. The expected average prices for consumers are presented as well. The comparisons of all three analyzed auction types with respect to optimal bids, expected costs, profits, and average prices conclude the analysis. Examples are given and supplement the analysis showing key results graphically.

4.1 Demand Ranges

A generator has all information except the marginal costs of his competitor in order to decide on his own bid. Because demand is known before the auction takes place, three demand ranges can be distinguished due to different competitive environments from the generator's point of view. The cases depend on the outcome for the winner and the loser.

Case A is a winner-takes-all situation. Just one generator is needed to match demand. The winner produces all electricity. His capacity utilization increases with demand. The loser does not sell anything because of his higher bid. Although electricity is a divisible commodity, the decision problem is the same as for an auction with an indivisible commodity.

In case B, both generators are necessary to serve demand but the loser does not fully utilize his production capacity. This weakens auctioneer's position in the market because she now depends on both generators. The winner utilizes his capacity fully. The loser can earn money even if he has offered the highest bid. Unused capacity is his only incentive to compete but he has less to lose compared with case A. This indicates a relaxed competitive environment for bidders and importantly influences their behavior.

S. Schöne, *Auctions in the Electricity Market*, Lecture Notes in Economics and Mathematical Systems 617,

The last case C refers to demands, which are equal or larger than the aggregated production capacity of both generators. It is the worst-case scenario for the auctioneer because both generators are needed with their full capacity. There is no remaining free production capacity. Demand may have to be rationed if it exceeds the market capacity.

The following definitions are helpful for clarification of the outcome determination and to simplify the analysis.

Definition 4.1.

(a) *The winner's outcome in an auction measured as the fraction of his production capacity is defined as*

$$\alpha = \min\left(\frac{d}{k}, 1\right), \quad \textit{with } \alpha \in (0,1]. \tag{4.1}$$

(b) *The fraction of production capacity the loser serves is given by*

$$\beta = \min\left(\frac{d}{k} - \alpha, 1\right), \quad \textit{with } \beta \in [0,1) \textit{ for } d < 2k \textit{ and } \beta = 1 \textit{ for } d \geq 2k. \tag{4.2}$$

(c) *The losing generator can serve the following proportion of the winner's outcome if demand does not reach full market production capacity. Taking into account the free production capacity due to the loser's obligations, the available fraction of production capacity is determined by*

$$\beta_f = \alpha - \beta, \quad \textit{with } \beta_f \in (0,1] \textit{ for } d < 2k \textit{ and } \beta_f = 0 \textit{ for } d \geq 2k. \tag{4.3}$$

The winner utilizes $\alpha \in (0,1]$ of his production capacity in case A. In all other cases, he sells as much as he can produce. The loser starts selling something in case B and utilizes $\beta \in (0,1)$ of his production capacity. This assured quantity reduces the competitive pressure of both generators. The incentive to compete in the auction is loser's free capacity $\beta_f \in (0,1)$. Competition vanishes along with the reduction of this free capacity.

4.2 Costs and Efficiency

Resources should be used economically. A society following this guidance is able to satisfy more needs either today or in the future. Almost all fuels of today's power plants are limited on earth. Its economic use helps to save resources for future generations. Another effect, which is important today, is the output reduction of carbon dioxide. This gas influences significantly the change of the climate. Its reduction may reduce or postpone negative effects of this change.

Before electricity markets were deregulated, monopolists were responsible for the electricity production and the safety of the grid. They had the incentive to utilize power plants with the lowest costs first in order to maximize their own profit. Their main concern was to minimize production costs given current demand. An economically desirable effect of using auctions is that auctions also lead to lowest production costs.

Theorem 4.1. *Suppose the value range of the used bid function is an element of the accepted bid range of the auctioneer. If the bid function used by generators is increasing with marginal costs, or demand is greater than or equal to production capacity of the whole market, the following statements are valid.*

(a) *The auction outcome provides production efficiency, i.e., the production schedule with the lowest costs.*

(b) *The expected production costs per electricity unit for $d \in (0,k]$ are given by*

$$E[c] = 2\int_{\underline{c}}^{\bar{c}} cF(c)\left[f(c|c) + F_i(c|c)\right]dc - 2\int_{\underline{c}}^{\bar{c}} F(c)\left[1 - F(c|c)\right]dc. \quad (4.4)$$

In the case of independence, the expression reduces to

$$E\left[c^i\right] = \bar{c} + \int_{\underline{c}}^{\bar{c}} F(c)^2\,dc - 2\int_{\underline{c}}^{\bar{c}} F(c)\,dc. \quad (4.5)$$

For demand $d \in (k, 2k)$, the expected cost is obtained by

$$\begin{aligned} E[c] = {} & 2\left[1 - \frac{k}{d}\right]\bar{c} - 2\frac{k}{d}\int_{\underline{c}}^{\bar{c}} F(c)\,dc \\ & + 2\left[\frac{2k}{d} - 1\right]\int_{\underline{c}}^{\bar{c}} F(c)\left\{c[f(c|c) + F_i(c|c)] + F(c|c)\right\}dc. \end{aligned} \quad (4.6)$$

Given independently distributed marginal costs, the expression reduces to

$$E\left[c^i\right] = \bar{c} + \left[\frac{2k}{d} - 1\right]\int_{\underline{c}}^{\bar{c}} F(c)^2\,dc - \frac{2k}{d}\int_{\underline{c}}^{\bar{c}} F(c)\,dc. \quad (4.7)$$

If demand equals or exceeds market production capacity $2k$, expected production costs are given by

$$E[c] = \bar{c} - \int_{\underline{c}}^{\bar{c}} F(c)\,dc. \quad (4.8)$$

Proof.

(a) In the case of using a bid function which is increasing in marginal costs, the generator with the lowest cost is always selected to serve the whole demand as long as his production capacity is not fully used. Only if demand is greater, the generator with higher costs is allowed to produce the remaining quantity. No other allocation of the right to produce power leads to a lower cost.

If demand is greater than or equal to the whole production capacity of both generators and the bids of both bidders are accepted by the auctioneer, the order of acceptance of the generators does not matter because both are needed with their entire capacity. Again, no other production schedule can be found that allows lower production costs. Hence, the production schedule, which is an outcome of the auction, provides the lowest cost and is therefore cost efficient.

(b) The expected cost per electricity unit is the expected average of total production costs. Note that the outcomes of the auctions would result in a production of α for the winner, see (4.1), and β for the loser, see (4.2). Presuming $j \neq i$, the formula is given by

$$E[c] = \frac{1}{\alpha+\beta} E\Big[\left\{\alpha c_i + \beta c_j\right\} 1_{\{b(c_i)<b(c_j)\}} + \left\{\beta c_i + \alpha c_j\right\} 1_{\{b(c_i)>b(c_j)\}}\Big].$$

This can be simplified to the following expression, using (4.3), because the bid functions are symmetric and increasing in c

$$\begin{aligned}
E[c] &= \frac{2}{\alpha+\beta} E\Big[\left\{\alpha c_i + \beta c_j\right\} 1_{\{c_i<c_j\}}\Big] \\
&= \frac{2}{\alpha+\beta} \int_{\underline{c}}^{\bar{c}} \int_{c}^{\bar{c}} [\alpha c + \beta \tau] f(\tau, c)\, d\tau dc \\
&= \frac{2\alpha}{\alpha+\beta} \int_{\underline{c}}^{\bar{c}} c f(c) [1 - F(c|c)]\, dc + \frac{2\beta}{\alpha+\beta} \int_{\underline{c}}^{\bar{c}} c f(c) F(c|c)\, dc \\
&= 2\frac{\alpha-\beta}{\alpha+\beta} \int_{\underline{c}}^{\bar{c}} c f(c) [1 - F(c|c)]\, dc + \frac{2\beta}{\alpha+\beta} \int_{\underline{c}}^{\bar{c}} c f(c)\, dc
\end{aligned} \quad (4.9)$$

$$= \frac{2\beta_f}{\alpha+\beta}\left\{ F(c)\,c[1-F(c|c)]\Big|_{\underline{c}}^{\bar{c}} + \int_{\underline{c}}^{\bar{c}} F(c)\,c[f(c|c)+F_i(c|c)]\,dc \right\}$$

$$-\frac{2\beta_f}{\alpha+\beta}\int_{\underline{c}}^{\bar{c}} F(c)\,[1-F(c|c)]\,dc$$

$$+\frac{2\beta}{\alpha+\beta}\left[cF(c)\Big|_{\underline{c}}^{\bar{c}} - \int_{\underline{c}}^{\bar{c}} F(c)\,dc \right]$$

$$= \frac{2\beta}{\alpha+\beta}\bar{c} + \frac{2\beta_f}{\alpha+\beta}\int_{\underline{c}}^{\bar{c}} F(c)\left\{ c[f(c|c)+F_i(c|c)] + F(c|c) \right\} dc$$

$$-\frac{2\alpha}{\alpha+\beta}\int_{\underline{c}}^{\bar{c}} F(c)\,dc. \tag{4.10}$$

Final expressions are given in the following taking into account (4.1), (4.2), and (4.3). The expected cost for $d \in (0,k]$ is obtained by

$$E[c] = 2\int_{\underline{c}}^{\bar{c}} cF(c)\,[f(c|c)+F_i(c|c)]\,dc - 2\int_{\underline{c}}^{\bar{c}} F(c)\,[1-F(c|c)]\,dc.$$

The expression for $d \in (k,2k)$ is given by

$$E[c] = 2\left[1-\frac{k}{d}\right]\bar{c} - 2\frac{k}{d}\int_{\underline{c}}^{\bar{c}} F(c)\,dc$$

$$+2\left[\frac{2k}{d}-1\right]\int_{\underline{c}}^{\bar{c}} F(c)\left\{ c[f(c|c)+F_i(c|c)] + F(c|c) \right\} dc.$$

For $d \geq 2k$, the expected cost is obtained by

$$E[c] = \bar{c} - \int_{\underline{c}}^{\bar{c}} F(c)\,dc.$$

If marginal costs are independently distributed, then expected costs, using (4.10), (3.11), (3.12), (3.14), and (3.3), are obtained by

$$E\left[c^i\right] = \frac{2\beta}{\alpha+\beta}\bar{c} + \frac{2\beta_f}{\alpha+\beta}\int_{\underline{c}}^{\bar{c}} F(c)\left[cf(c)+F(c)\right]dc - \frac{2\alpha}{\alpha+\beta}\int_{\underline{c}}^{\bar{c}} F(c)\,dc.$$

For simplicity, the following integral is solved separately

$$\begin{aligned}\int_{\underline{c}}^{\bar{c}} cF(c)f(c)\,dc &= cF(c)^2\Big|_{\underline{c}}^{\bar{c}} - \int_{\underline{c}}^{\bar{c}} F(c)\left[F(c)+cf(c)\right]dc \\ &= \bar{c} - \int_{\underline{c}}^{\bar{c}} F(c)^2\,dc - \int_{\underline{c}}^{\bar{c}} cF(c)f(c)\,dc \\ &= \frac{1}{2}\left[\bar{c} - \int_{\underline{c}}^{\bar{c}} F(c)^2\,dc\right].\end{aligned} \tag{4.11}$$

This expression and the definitions of β_f, see (4.3), are used to get

$$\begin{aligned}E\left[c^i\right] &= \frac{2\beta}{\alpha+\beta}\bar{c} - \frac{2\alpha}{\alpha+\beta}\int_{\underline{c}}^{\bar{c}} F(c)\,dc + \frac{2\beta_f}{\alpha+\beta}\int_{\underline{c}}^{\bar{c}} F(c)^2\,dc \\ &\quad + \frac{\beta_f}{\alpha+\beta}\left[\bar{c} - \int_{\underline{c}}^{\bar{c}} F(c)^2\,dc\right] \\ &= \bar{c} + \frac{\alpha-\beta}{\alpha+\beta}\int_{\underline{c}}^{\bar{c}} F(c)^2\,dc - \frac{2\alpha}{\alpha+\beta}\int_{\underline{c}}^{\bar{c}} F(c)\,dc.\end{aligned}$$

The final expression for independently distributed marginal costs is obtained by, using (4.1) and (4.2)

$$E\left[c^i\right] = \begin{cases} \bar{c} + \int_{\underline{c}}^{\bar{c}} F(c)^2\,dc - 2\int_{\underline{c}}^{\bar{c}} F(c)\,dc & : d \in (0,k], \\ \bar{c} + \left[\frac{2k}{d} - 1\right]\int_{\underline{c}}^{\bar{c}} F(c)^2\,dc - \frac{2k}{d}\int_{\underline{c}}^{\bar{c}} F(c)\,dc & : d \in (k,2k), \\ \bar{c} - \int_{\underline{c}}^{\bar{c}} F(c)\,dc & : d \geq 2k. \end{cases}$$

□

It is an economically reasonable assumption that bids rise with higher costs. The assumption is based on the idea that competitive advantage alone is the source of success in a market. Lowest costs are an example for competitive advantage. As long as this assumption holds true, the auctioneer selects the generator with the lowest marginal costs by choosing the lowest bid. Hence, electricity is produced at the lowest cost as stated in Theorem 4.1 (a).

An interesting demand range is given with $d \in (k, 2k)$. In this case, the less efficient generator produces an increasing share of electricity as demand rises. This aspect should be found in the formula.

Lemma 4.1. *Expected production costs for demand range $d \in (k, 2k)$ are a linear combination of costs for $d \in (0, k]$ and $d > 2k$.*

$$E\left[c \,|\, d \in (k,2k)\right] = \lambda E\left[c \,|\, d \in (0,k]\right] + [1-\lambda] E\left[c \,|\, d \geq 2k\right], \tag{4.12}$$

where λ is given by

$$\lambda = \tfrac{2k}{d} - 1. \tag{4.13}$$

Proof.

The right term of (4.12) is given by, using (4.4), (4.8), and (4.13)

$$\lambda E\left[c \,|\, d \in (0,k]\right] + [1-\lambda] E\left[c \,|\, d \geq 2k\right]$$

$$= 2\left[\frac{2k}{d} - 1\right] \int_{\underline{c}}^{\bar{c}} F(c) \left\{ c[f(c|c) + F_i(c|c)] - 1 + F(c|c) \right\} dc$$
$$+ 2\left[1 - \frac{k}{d}\right] \left[\bar{c} - \int_{\underline{c}}^{\bar{c}} F(c)\, dc\right]$$

$$= 2\left[\frac{2k}{d} - 1\right] \int_{\underline{c}}^{\bar{c}} F(c) \left\{ c[f(c|c) + F_i(c|c)] + F(c|c) \right\} dc - 2\frac{k}{d} \int_{\underline{c}}^{\bar{c}} F(c)\, dc$$
$$+ 2\left[1 - \frac{k}{d}\right] \int_{\underline{c}}^{\bar{c}} F(c)\, dc + 2\left[1 - \frac{k}{d}\right] \left[\bar{c} - \int_{\underline{c}}^{\bar{c}} F(c)\, dc\right]$$

$$= 2\left[1-\frac{k}{d}\right]\bar{c} - 2\frac{k}{d}\int_{\underline{c}}^{\bar{c}} F(c)\,dc$$

$$+2\left[\frac{2k}{d}-1\right]\int_{\underline{c}}^{\bar{c}} F(c)\left\{c[f(c|c)+F_i(c|c)]+F(c|c)\right\}dc.$$

It results in the expression for expected costs with $d \in (k, 2k)$. □

Remark 4.1. *Note that $\lambda \in [0,1]$ holds true.*

Example 4.1. *Suppose the probability distribution given by Definition 3.5.*

(a) *The expected production costs per electricity unit for $d \in (0,k]$ are given by*

$$E[c] = \begin{cases} \frac{1}{3} & : a = 0 \\ \frac{4}{5}\left[\frac{2}{3} - \frac{1}{4+a}\right] & : a > 0. \end{cases} \tag{4.14}$$

(b) *The expected production costs per electricity unit for $d \in (k,2k)$ are obtained by*

$$E[c] = \begin{cases} \frac{1}{3}\left[2-\frac{k}{d}\right] & : a = 0 \\ \frac{4}{15}\left[3-\frac{2}{4+a}\right] - \frac{4k}{15d}\left[1+\frac{1}{4+a}\right] & : a > 0. \end{cases} \tag{4.15}$$

(c) *The expected production costs per electricity unit for $d \geq 2k$ are given by*

$$E[c] = \begin{cases} \frac{1}{2} & : a = 0 \\ \frac{2}{3}\left[1-\frac{1}{4+a}\right] & : a > 0. \end{cases} \tag{4.16}$$

Proof.

(a) The expected production cost per electricity unit for independently distributed marginal costs, using (4.5) and (3.25), are given by

$$E\left[c^i\right] = 1 + \int_0^1 c^2\,dc - 2\int_0^1 c\,dc$$

$$= 1 + \left.\frac{c^3}{3}\right|_0^1 - \left.c^2\right|_0^1 = \frac{1}{3}.$$

The expected production costs for strong affiliation are derived using following term, see (3.25), (3.26), and (3.32)

$$\int_0^1 cF(c)\left[f(c|c)+F_i(c|c)\right]dc$$

$$=\int_0^1 \frac{1+\frac{a}{4}c}{1+\frac{a}{4}}c^2\left\{\frac{1+ac^2}{1+\frac{a}{2}c}-\frac{a[1-c]c}{2\left[1+\frac{a}{2}c\right]^2}\right\}dc$$

$$=\frac{2}{4+a}\int_0^1 [4+ac]\frac{[2+ac][1+ac^2]-a[1-c]c}{[2+ac]^2}c^2\,dc$$

$$=\frac{2}{4+a}\int_0^1 [4+ac]\frac{a^2c^3+3ac^2+2}{[2+ac]^2}c^2\,dc$$

$$=\frac{2}{4+a}\int_0^1 [1+ac]\frac{4+ac}{2+ac}c^3\,dc+\frac{4}{4+a}\int_0^1 \frac{4+ac}{[2+ac]^2}[1-c]c^2\,dc$$

$$=\frac{2}{4+a}\int_0^1 [1+ac]c^3\,dc+\frac{4}{4+a}\int_0^1 \frac{1+ac}{2+ac}c^3\,dc+\frac{4}{4+a}\int_0^1 \frac{1-c}{2+ac}c^2\,dc$$

$$+\frac{8}{4+a}\int_0^1 \frac{1-c}{[2+ac]^2}c^2\,dc$$

$$=\frac{2}{4+a}\int_0^1 [1+ac]c^3\,dc+\frac{4}{4+a}\int_0^1 c^3\,dc-\frac{4}{4+a}\int_0^1 \frac{c^3}{2+ac}dc$$

$$+\frac{4}{4+a}\int_0^1 \frac{1-c}{2+ac}c^2\,dc+\frac{8}{4+a}\int_0^1 \frac{1-c}{[2+ac]^2}c^2\,dc$$

$$=\frac{2}{4+a}\int_0^1 \frac{a^2c^3+5ac^2+2c+2}{2+ac}c^2\,dc+\frac{8}{4+a}\int_0^1 \frac{1-c}{[2+ac]^2}c^2\,dc$$

$$=\frac{2}{4+a}\int_0^1 [ac+3]c^3\,dc+\frac{4}{4+a}\int_0^1 \frac{1-2c}{2+ac}c^2\,dc+\frac{8}{4+a}\int_0^1 \frac{1-c}{[2+ac]^2}c^2\,dc.$$

The term for strong affiliation with $a > 0$ is determined by

$$\frac{2}{4+a}\int_0^1 [ac+3]c^3\,dc + \frac{4}{4+a}\int_0^1 \frac{1-2c}{2+ac}c^2\,dc + \frac{32}{a^3}\frac{2+a}{4+a}\int_0^1 \frac{1}{[2+ac]^2}\,dc$$

$$-\frac{8}{a^3[4+a]}\int_0^1 \frac{a^2c^2-a^2c-2ac+2a+4}{2+ac}\,dc$$

$$= \frac{2}{4+a}\int_0^1 [ac+3]c^3\,dc + \frac{4}{a[4+a]}\int_0^1 [1-2c]c\,dc + \frac{8}{a^2[4+a]}\int_0^1 c\,dc$$

$$-\frac{16}{a^3}\frac{2+a}{4+a}\int_0^1 \frac{1}{2+ac}\,dc + \frac{32}{a^3}\frac{2+a}{4+a}\int_0^1 \frac{1}{[2+ac]^2}\,dc$$

$$= \frac{2}{5} - \frac{1}{10[4+a]} - \frac{2}{3a[4+a]} + \frac{4}{a^2[4+a]} + \frac{16}{a^3[4+a]}$$

$$-\frac{16[2+a]}{a^4[4+a]}\ln\left(1+\frac{a}{2}\right).$$

$$= \frac{2}{5} - \frac{1}{4+a}\left[\frac{1}{10} + \frac{2}{3a} - \frac{4}{a^2} - \frac{16}{a^3} + 16\frac{2+a}{a^4}\ln\left(1+\frac{a}{2}\right)\right]. \tag{4.17}$$

The expected production costs for $a > 0$, using (4.4), (3.25), (3.27), and (4.17), are obtained by

$$E[c] = \frac{4}{5} - \frac{2}{4+a}\left[\frac{1}{10} + \frac{2}{3a} - \frac{4}{a^2} - \frac{16}{a^3} + 16\frac{2+a}{a^4}\ln\left(1+\frac{a}{2}\right)\right]$$

$$-2\int_0^1 c\,\frac{1+\frac{a}{4}c}{1+\frac{a}{4}}\left[1-\frac{1+\frac{a}{2}c^2}{1+\frac{a}{2}c}c\right]dc$$

$$= \frac{4}{5} - \frac{2}{4+a}\left[\frac{1}{10} + \frac{2}{3a} - \frac{4}{a^2} - \frac{16}{a^3} + 16\frac{2+a}{a^4}\ln\left(1+\frac{a}{2}\right)\right]$$

$$+\frac{32}{a^3}\frac{2+a}{4+a}\int_0^1 \frac{1}{2+ac}\,dc + \frac{2}{4+a}\int_0^1 [ac^4-ac^2+2c^3+2c^2-4c]\,dc$$

$$-\frac{8}{a^3[4+a]}\int_0^1 [a^2c^2-a^2c-2ac+2a+4]\,dc$$

$$= \frac{4}{5} - \frac{2}{4+a}\left[\frac{1}{10} + \frac{2}{3a} - \frac{4}{a^2} - \frac{16}{a^3} + 16\frac{2+a}{a^4}\ln\left(1+\frac{a}{2}\right)\right]$$

$$+\frac{32}{a^4}\frac{2+a}{4+a}\ln\left(1+\frac{a}{2}\right)$$

$$+\frac{2}{4+a}\left[\frac{a}{5}c^5 - \frac{a}{3}c^3 + \frac{1}{2}c^4 + \frac{2}{3}c^3 - 2c^2\right]\Bigg|_0^1$$

$$-\frac{8}{a^3[4+a]}\left[\frac{a^2}{3}c^3 - \frac{a^2}{2}c^2 - ac^2 + 2ac + 4c\right]\Bigg|_0^1$$

$$= \frac{4}{5} - \frac{2}{4+a}\left[\frac{1}{10} + \frac{2}{3a} - \frac{4}{a^2} - \frac{16}{a^3} + 16\frac{2+a}{a^4}\ln\left(1+\frac{a}{2}\right)\right]$$

$$-\frac{4}{15} - \frac{1}{4+a}\left[\frac{3}{5} - \frac{4}{3a} + \frac{8}{a^2} + \frac{32}{a^3} - 32\frac{2+a}{a^4}\ln\left(1+\frac{a}{2}\right)\right]$$

$$= \frac{4}{5}\left[\frac{2}{3} - \frac{1}{4+a}\right].$$

(b) Given independently distributed marginal costs, the expected production costs per electricity unit, using (4.7) and (3.25), are obtained by

$$E\left[c^i\right] = 1 + \left[\frac{2k}{d} - 1\right]\int_0^1 c^2\,dc - \frac{2k}{d}\int_0^1 c\,dc$$

$$= 1 + \left[\frac{2k}{d} - 1\right]\frac{c^3}{3}\Bigg|_0^1 - \frac{k}{d}c^2\Bigg|_0^1$$

$$= \frac{1}{3}\left[2 - \frac{k}{d}\right].$$

Taking into account (4.6), the same expression but for strong affiliation with $a > 0$ is given by

$$E[c] = 2\left[1 - \frac{k}{d}\right] - \frac{2k}{d}\int_0^1 F(c)\,dc + 2\left[\frac{2k}{d} - 1\right]\int_0^1 F(c)\,F(c|\,c)\,dc$$

$$+2\left[\frac{2k}{d} - 1\right]\int_0^1 cF(c)\left[f(c|\,c) + F_i(c|\,c)\right]dc$$

$$= 2\left[1-\frac{k}{d}\right]\left[1-\int_0^1 F(c)\,dc\right]+2\left[\frac{2k}{d}-1\right]$$
$$\times\left\{\int_0^1 cF(c)\left[f(c|c)+F_i(c|c)\right]dc-\int_0^1 F(c)\left[1-F(c|c)\right]dc\right\}.$$

The last summand contains the expression of expected production costs for $d \in (0,k]$. Hence, the equation, using (4.14) and (3.25), is rewritten to

$$E[c] = 2\left[1-\frac{k}{d}\right]\left[1-\int_0^1 \frac{4+ac}{4+a}\,c\,dc\right]+\frac{4}{5}\left[\frac{2k}{d}-1\right]\left[\frac{2}{3}-\frac{1}{4+a}\right]$$
$$= 2\left[1-\frac{k}{d}\right]\left\{1-\frac{1}{4+a}\left[2c^2+\frac{a}{3}c^3\right]\Big|_0^1\right\}+\frac{4}{5}\left[\frac{2k}{d}-1\right]\left[\frac{2}{3}-\frac{1}{4+a}\right]$$
$$= \frac{4}{3}\left[1-\frac{k}{d}\right]\frac{3+a}{4+a}+\frac{4}{5}\left[\frac{2k}{d}-1\right]\left[\frac{2}{3}-\frac{1}{4+a}\right]$$
$$= \frac{8k}{5d}\left[\frac{2}{3}-\frac{1}{4+a}\right]-\frac{4}{3}\left[\frac{k}{d}-1\right]\frac{3+a}{4+a}-\frac{4}{5}\left[\frac{2}{3}-\frac{1}{4+a}\right]$$
$$= \frac{4k}{15d}\left[4-\frac{21+5a}{4+a}\right]-\frac{4}{15}\left[2-\frac{18+5a}{4+a}\right]$$
$$= \frac{4}{15}\left[3-\frac{2}{4+a}\right]-\frac{4k}{15d}\left[1+\frac{1}{4+a}\right].$$

(c) The expected production costs per electricity unit, using (4.8) and (3.25), are given by

$$E[c] = 1-\int_0^1 \frac{1+\frac{a}{4}c}{1+\frac{a}{4}}\,c\,dc$$
$$= 1-\frac{2}{4+a}\,c^2\Big|_0^1-\frac{a}{3[4+a]}\,c^3\Big|_0^1=\frac{2}{3}\left[1-\frac{1}{4+a}\right].$$

The expression for independently distributed marginal costs with $a=0$ reduces to

$$E\left[c^i\right]=\frac{1}{2}.$$

□

Figure 4.1 shows the expected costs of the example. The graph for independent marginal costs is dominated by graphs for strongly affiliated marginal costs. Strong affiliation results in a higher probability that the second generator has high costs when the first generator has high cost too. This shift of probabilities increases the expected costs. This can be seen by comparing the second graph ($a = 1$) with the third graph ($a = 4$). The stronger the affiliation of marginal costs is, measured by greater value of a, the higher the expected costs. This is reflected by the dominance of the last graph.

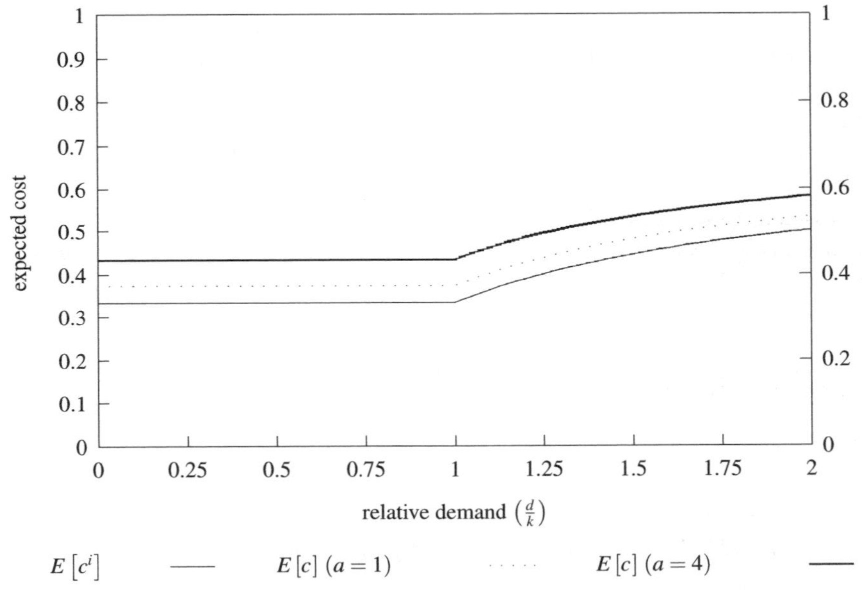

Fig. 4.1 Expected cost per electricity unit given production efficiency

In case A with $d \in (0,k]$, expected costs are independent of demand d and production capacity k. Just one generator is sufficient to serve demand. Hence, only his marginal costs, which is constant, is relevant. Costs of case C, with $d \geq 2k$, also do not depend on demand or production capacity. The reason here is that both generators have the same production capacity and utilize it fully. Hence, they act like one big generator with average costs of the winner and the loser.

Case B, with $d \in (k,2k)$, lies in the middle between case A and case C. Now the loser produces something but does not fully utilize his capacity. The loser's proportion of produced electricity increases with higher demand. Accordingly, the winner's fraction decreases. Hence, the higher cost of the loser becomes more important, which raises the expected average costs. This mechanism is recognizable in (4.12). λ, defined by (4.13), represents the loser's unused production capacity relative to demand or missed opportunity. On the other hand, $[1-\lambda]$ stands for the loser's utilized capacity. The expected average cost increases linearly with rising demand because the loser serves marginal demand. His cost disadvantage is constant for every produced electricity unit. Factor $[1-\lambda]$ determines how much the disadvantage raises average costs.

4.3 Discriminatory Auction

4.3.1 Case A: A Single Generator Serves the Demand

Scheduled generators get their own bids paid for each electricity unit sold. This pricing rule induces to bid higher than marginal costs. Otherwise, they would not make any profit. All bids should therefore show bid shading, a positive difference between bid and cost.

Theorem 4.2. *Suppose $d \in (0,k]$. The unique Bayes–Nash equilibrium in symmetric and strictly increasing bidding strategies is given by $\left(b_d^*, b_d^*\right)$ with the following bid function*

$$b_d^*(c) = \left[\bar{p} - \bar{c} + \int_c^{\bar{c}} \exp\left(\int_\tau^{\bar{c}} \frac{f(t|t)}{1-F(t|t)}\,dt\right) d\tau\right] \exp\left(-\int_c^{\bar{c}} \frac{f(t|t)}{1-F(t|t)}\,dt\right) + c. \tag{4.18}$$

The optimal bid function for independent distribution reduces to

$$b_d^{*i}(c) = \begin{cases} c + \dfrac{\bar{c} - c - \int_c^{\bar{c}} F(\tau)\,d\tau}{1-F(c)} & : c \in [\underline{c}, \bar{c}) \\ \bar{p} & : c = \bar{c}. \end{cases} \tag{4.19}$$

Proof.

The proof is attained in four steps. Firstly, the bid function is found which maximizes the expected profit. Secondly, it is shown that the auction really takes place. Thirdly, it is proved that a Bayes–Nash equilibrium is obtained by the bid function, and finally that only one Bayes–Nash equilibrium exists. Due to the symmetry of the generators, the proof is reduced to the analysis of generator i.

The decision problem for $d \in (k, 2k)$ can be analyzed in the same way as for $d \in (0, k]$. The quantity sold by the winner is simply denoted by α, see (4.1). The quantity sold by the loser is referred to as β, see (4.2). The proof is already given for the wider demand range.

Step 1: Bid Function

The expected profit of generator i for $j \neq i$, using (4.1), (4.2), and presuming a strictly increasing bid function, is

$$\begin{aligned}\pi_i(b_i, c_i) &= kE\left[\alpha[b_i - c_i]1_{\{b_i<b_j\}} + \beta[b_i - c_i]1_{\{b_i>b_j\}} \middle| C_i = c_i\right] \\ &= k[b_i - c_i]\left\{\alpha - [\alpha - \beta]E\left[1_{\{b_i>b_j\}} \middle| C_i = c_i\right]\right\} \\ &= k[b_i - c_i]\left[\alpha - \beta_f \int\limits_{\underline{c}}^{b^{-1}(b_i)} f(\tau|c_i)\, d\tau\right].\end{aligned}$$

b_i denotes the chosen bid. The third factor is often used. Hence, following function is introduced for simplicity

$$\begin{aligned}y(c_j|c_i) &= \alpha - \beta_f \int\limits_{\underline{c}}^{c_j} f(\tau|c_i)\, d\tau \\ &= \alpha - \beta_f F(c_j|c_i). \end{aligned} \tag{4.20}$$

$y(c|c)$ can be interpreted as the expected degree of production capacity utilization given marginal costs c.[1] The function has following properties, see (4.3)

$$\begin{aligned} y(c_j|c_i) &= 0, \quad \text{for } \beta = 0 \text{ and } c_j = \bar{c}, \\ y(c_j|c_i) &> 0, \quad \text{otherwise}. \end{aligned} \tag{4.21}$$

[1] The interpretation holds true if the bid function does increase with marginal costs. The validity of this assumption is shown later. A precise interpretation is that $y(c_j|c_i)$ represents the fraction of production capacity that generator i can expect to utilize knowing his costs c_i and bidding such that his bid is ordered like c_j among marginal costs. The rewritten function (4.20) may make it clearer $y(c_j|c_i) = \alpha[1 - F(c_j|c_i)] + \beta F(c_j|c_i)$.

If the loser gets nothing, which is represented by $\beta = 0$ according to (4.3), β_f becomes α. Hence, (4.20) is obtained by

$$y(c_j|c_i) = \alpha[1 - F(c_j|c_i)].$$

The term is zero for $c_j = \bar{c}$, and otherwise positive because $\alpha > 0$, due to (4.1), and $F(c_j|c_i) < 1$, $\forall c_j \in [\underline{c}, \bar{c})$. The function is determined in all other cases by

$$y(c_j|c_i) = \alpha[1 - F(c_j|c_i)] + \beta F(c_j|c_i).$$

The function is always positive because $F(c_j|c_i)$ is limited to one. Hence, at least one summand is positive due to positive values for α and β, see (4.1) and (4.2).

The expected profit is now given by, using (4.20)

$$\pi_i(b_i, c_i) = k[b_i - c_i]y\left(b^{-1}(b_i)|c_i\right). \tag{4.22}$$

The first derivative of the expected profit is obtained by

$$\frac{\partial \pi_i(b_i, c_i)}{\partial b_i} = k\left[y\left(b^{-1}(b_i)|c_i\right) + \frac{b_i - c_i}{b\left(b^{-1}(b_i)\right)'} y_j\left(b^{-1}(b_i)|c_i\right)\right] \tag{4.23}$$

using $b(\cdot)'$ and $y_j(\cdot|\cdot)$. The first function, $b(c)'$, denotes the first derivative of the bid function with respect to marginal costs c. The second is the first derivative of (4.20) with respect to c_j

$$\begin{aligned} y_j(c_j|c_i) &\equiv \frac{\partial y(c_j|c_i)}{\partial c_j} \\ &= (-)\beta_f f(c_j|c_i). \end{aligned} \tag{4.24}$$

This derivative has following property

$$y_j(c_j|c_i) < 0. \tag{4.25}$$

The reason is that β_f is always positive for $d < 2k$, due to (4.3). Additionally, $f(c_j|c_i)$ is positive because all combinations of marginal costs are presumed to be feasible.

The optimal bid function is obtained using the necessary condition of an optimum $\frac{\partial \pi_i(b_i,c_i)}{\partial b_i} \stackrel{!}{=} 0$. This is equivalent to, see (4.23)

$$y\left(b^{-1}(b_i)\,|\,c_i\right) + \frac{b_i - c_i}{b\left(b^{-1}(b_i)\right)'}\, y_j\left(b^{-1}(b_i)\,|\,c_i\right) = 0.$$

$b_i = b(c_i)$ has to be valid for a symmetric Bayes–Nash equilibrium. Hence, the profit maximum is found if the following holds true

$$b(c)' + b(c)\,\frac{y_j(c|\,c)}{y(c|\,c)} = c\,\frac{y_j(c|\,c)}{y(c|\,c)}. \tag{4.26}$$

A solution of the linear non-homogeneous differential equation is[2]

$$b(c) = b_0\, \mathrm{e}^{\int_\theta^c \frac{(-)y_j(t|t)}{y(t|t)}dt} + \int_\theta^c \tau\, \frac{y_j(\tau|\,\tau)}{y(\tau|\,\tau)}\, \mathrm{e}^{\int_\tau^c \frac{(-)y_j(t|t)}{y(t|t)}dt}\, d\tau. \tag{4.27}$$

The upper limit of the bid function is set by $\hat{b}$. Because the generator with the highest marginal cost must earn a profit, this limit is greater than $\bar{c}$. It does not exceed reservation price $\bar{p}$ because otherwise the bid is not accepted according to the auction rules. Hence, the limit is given by

$$\hat{b} \in (\bar{c}, \bar{p}]. \tag{4.28}$$

The generator with the highest marginal costs sets the upper limit because a strictly increasing bid function is assumed. Hence, the bid function given marginal costs $\bar{c}$ is obtained by

$$b(\bar{c}) = \hat{b} \tag{4.29}$$

$$\Longleftrightarrow$$

$$\hat{b} \quad = b_0\, \mathrm{e}^{\int_\theta^{\bar{c}} \frac{(-)y_j(t|t)}{y(t|t)}dt} + \int_\theta^{\bar{c}} \tau\, \frac{y_j(\tau|\,\tau)}{y(\tau|\,\tau)}\, \mathrm{e}^{\int_\tau^{\bar{c}} \frac{(-)y_j(t|t)}{y(t|t)}dt}\, d\tau$$

$$\Longleftrightarrow$$

$$b_0 \quad = \left[\hat{b} - \int_\theta^{\bar{c}} \tau\, \frac{y_j(\tau|\,\tau)}{y(\tau|\,\tau)}\, \mathrm{e}^{\int_\tau^{\bar{c}} \frac{(-)y_j(t|t)}{y(t|t)}dt}\, d\tau\right] \mathrm{e}^{\int_\theta^{\bar{c}} \frac{y_j(t|t)}{y(t|t)}dt}.$$

[2] For a general solution of linear first-order differential equations see Sydsæter, Strøm, and Berck (1999), p. 62.

Substituting b_0, the bid function (4.27) is given by

$$b(c) = \left[\hat{b} - \int_{\theta}^{\bar{c}} \tau \frac{y_j(\tau|\tau)}{y(\tau|\tau)} \, e^{\int_{\tau}^{\bar{c}} \frac{(-)y_j(t|t)}{y(t|t)} dt} d\tau\right] e^{\int_{c}^{\bar{c}} \frac{y_j(t|t)}{y(t|t)} dt}$$

$$+ \int_{\theta}^{c} \tau \frac{y_j(\tau|\tau)}{y(\tau|\tau)} \, e^{\int_{\tau}^{c} \frac{(-)y_j(t|t)}{y(t|t)} dt} d\tau$$

$$= \hat{b} e^{\int_{c}^{\bar{c}} \frac{y_j(t|t)}{y(t|t)} dt} - \int_{\theta}^{\bar{c}} \tau \frac{y_j(\tau|\tau)}{y(\tau|\tau)} \, e^{\int_{\tau}^{c} \frac{(-)y_j(t|t)}{y(t|t)} dt} d\tau + \int_{\theta}^{c} \tau \frac{y_j(\tau|\tau)}{y(\tau|\tau)} \, e^{\int_{\tau}^{c} \frac{(-)y_j(t|t)}{y(t|t)} dt} d\tau$$

$$= \hat{b} e^{\int_{c}^{\bar{c}} \frac{y_j(t|t)}{y(t|t)} dt} - \int_{c}^{\bar{c}} \tau \frac{y_j(\tau|\tau)}{y(\tau|\tau)} \, e^{\int_{c}^{\tau} \frac{y_j(t|t)}{y(t|t)} dt} d\tau$$

$$= \hat{b} e^{\int_{c}^{\bar{c}} \frac{y_j(t|t)}{y(t|t)} dt} - \tau e^{\int_{c}^{\tau} \frac{y_j(t|t)}{y(t|t)} dt} \Bigg|_{c}^{\bar{c}} + \int_{c}^{\bar{c}} e^{\int_{c}^{\tau} \frac{y_j(t|t)}{y(t|t)} dt} d\tau$$

$$= \left[\hat{b} - \bar{c}\right] e^{\int_{c}^{\bar{c}} \frac{y_j(t|t)}{y(t|t)} dt} + \int_{c}^{\bar{c}} e^{\int_{c}^{\tau} \frac{y_j(t|t)}{y(t|t)} dt} d\tau + c$$

$$= \left[\hat{b} - \bar{c} + \int_{c}^{\bar{c}} e^{\int_{\tau}^{\bar{c}} \frac{(-)y_j(t|t)}{y(t|t)} dt} d\tau\right] e^{\int_{c}^{\bar{c}} \frac{y_j(t|t)}{y(t|t)} dt} + c. \tag{4.30}$$

The first derivative of the bid function, using (4.26), is given by

$$\frac{\partial b(c)}{\partial c} = (-)\left[b(c) - c\right] \frac{y_j(c|c)}{y(c|c)}. \tag{4.31}$$

The bid function was presumed to be increasing strictly. Hence, the following condition has to be valid

$$\frac{\partial b(c)}{\partial c} > 0, \quad \forall\, c \in [\underline{c}, \bar{c}]. \tag{4.32}$$

This is equivalent to, using (4.31)

$$\left[b(c) - c\right] \frac{y_j(c|c)}{y(c|c)} < 0.$$

Because $y_j(c|c)$ is always negative due to (4.25) and $y(c|c)$ is non-negative, see (4.21), the validity of this inequation depends on the sign of the first term. Therefore, the following has to hold true

$$b(c) > c. \tag{4.33}$$

This is equivalent to, using bid function (4.30)

$$\left[\hat{b} - \bar{c} + \int_{c}^{\bar{c}} e^{\int_{\tau}^{\bar{c}} \frac{(-)y_j(t|t)}{y(t|t)} dt} d\tau\right] e^{\int_{c}^{\bar{c}} \frac{y_j(t|t)}{y(t|t)} dt} > 0$$

$$\Longleftrightarrow$$

$$\hat{b} > \bar{c} - \int_{c}^{\bar{c}} e^{\int_{\tau}^{\bar{c}} \frac{(-)y_j(t|t)}{y(t|t)} dt} d\tau.$$

The right term of the inequation does not exceed $\bar{c}$ because the upper limit of marginal costs is $\bar{c}$ and the integral will not become negative. Hence, the condition holds true if and only if $\hat{b} > \bar{c}$. This was assumed by (4.28) because the highest accepted bid $\bar{p}$ is higher than the highest marginal costs $\bar{c}$. Therefore, inequation (4.33) is valid and the bid function is indeed strictly increasing with respect to c, which was a necessary condition of finding the function itself.

The following derivatives are needed to check that the bid function (4.30) maximizes the profit. Firstly, derivatives of the auxiliary function (4.20) are given. The second derivative of the bid function ends the list.

The first derivative of $y(c_j|c_i)$ with respect to c_i, using (3.8), is obtained by

$$\begin{aligned} y_i(c_j|c_i) &\equiv \frac{\partial y(c_j|c_i)}{\partial c_i} \\ &= (-)\beta_f F_i(c_j|c_i). \end{aligned} \tag{4.34}$$

The second derivative of $y(c_j|c_i)$ with respect to c_i and c_j is derived from (4.24) using (3.7)

$$\begin{aligned} y_{ji}(c_j|c_i) &\equiv \frac{\partial^2 y(c_j|c_i)}{\partial c_j \partial c_i} \\ &= (-)\beta_f \frac{\partial f(c_j|c_i)}{\partial c_i} = (-)\beta_f f_i(c_j|c_i). \end{aligned} \tag{4.35}$$

The second derivative of $y(c_j|c_i)$ with respect to c_j is denoted by

$$y_{jj}(c_j|c_i) \equiv \frac{\partial^2 y(c_j|c_i)}{\partial c_j^2}. \tag{4.36}$$

The second derivative of the bid function, using (4.31), is given by

$$\begin{aligned}
\frac{\partial^2 b(c)}{\partial c^2} &= (-)\left[b(c)' - 1\right]\frac{y_j(c|c)}{y(c|c)} - [b(c) - c]\frac{y_{jj}(c|c) + y_{ji}(c|c)}{y(c|c)} \\
&\quad + [b(c) - c]\frac{y_j(c|c)\,[y_j(c|c) + y_i(c|c)]}{y(c|c)^2} \\
&= \frac{y_j(c|c)}{y(c|c)} - b(c)'\frac{y_j(c|c)}{y(c|c)} + b(c)'\frac{y_{jj}(c|c) + y_{ji}(c|c)}{y_j(c|c)} \\
&\quad - b(c)'\frac{y_j(c|c) + y_i(c|c)}{y(c|c)} \\
&= \frac{y_j(c|c)}{y(c|c)} \\
&\quad + b(c)'\left[\frac{y_{jj}(c|c) + y_{ji}(c|c)}{y_j(c|c)} - \frac{2y_j(c|c) + y_i(c|c)}{y(c|c)}\right].
\end{aligned} \tag{4.37}$$

The bid function only ensures an expected profit maximum if the second derivative of the profit function is negative. Using (4.23), this derivative is obtained by

$$\begin{aligned}
\frac{\partial^2 \pi_i(b_i, c_i)}{\partial b_i^2} &= k\left[\frac{y_j\left(b^{-1}(b_i)\,|\,c_i\right)}{b\left(b^{-1}(b_i)\right)'} + \frac{y_j\left(b^{-1}(b_i)\,|\,c_i\right)}{b\left(b^{-1}(b_i)\right)'}\right. \\
&\quad + \frac{b_i - c_i}{b\left(b^{-1}(b_i)\right)'}\,\frac{y_{jj}\left(b^{-1}(b_i)\,|\,c_i\right)}{b\left(b^{-1}(b_i)\right)'} \\
&\quad \left. - \frac{b_i - c_i}{\left[b\left(b^{-1}(b_i)\right)'\right]^2}\,\frac{b\left(b^{-1}(b_i)\right)''}{b\left(b^{-1}(b_i)\right)'}\,y_j\left(b^{-1}(b_i)\,|\,c_i\right)\right]
\end{aligned}$$

$$= k\left\{2\frac{y_j\left(b^{-1}(b_i)|c_i\right)}{b\left(b^{-1}(b_i)\right)'}+\frac{b_i-c_i}{\left[b\left(b^{-1}(b_i)\right)'\right]^2}\right.$$
$$\left.\times\left[y_{jj}\left(b^{-1}(b_i)|c_i\right)-\frac{b\left(b^{-1}(b_i)\right)''}{b\left(b^{-1}(b_i)\right)'}y_j\left(b^{-1}(b_i)|c_i\right)\right]\right\}.$$

In Bayes–Nash equilibrium, the bid is calculated according to the bid function (4.30). The derivative of the profit function, using (4.31) and (4.37), is given by

$$\frac{\partial^2\pi(b(c),c)}{\partial b(c)^2}$$
$$= k\left\{2\frac{y_j(c|c)}{b(c)'}+\frac{b(c)-c}{[b(c)']^2}\left[y_{jj}(c|c)-b(c)''\frac{y_j(c|c)}{b(c)'}\right]\right\}$$
$$= k\left\{2\frac{y_j(c|c)}{b(c)'}-\frac{y(c|c)y_{jj}(c|c)}{b(c)'y_j(c|c)}+b(c)''\frac{y(c|c)}{[b(c)']^2}\right\}$$
$$= k\left\{2\frac{y_j(c|c)}{b(c)'}-\frac{y(c|c)y_{jj}(c|c)}{b(c)'y_j(c|c)}+\frac{y_j(c|c)}{[b(c)']^2}+y(c|c)\frac{y_{jj}(c|c)+y_{ji}(c|c)}{b(c)'y_j(c|c)}\right.$$
$$\left.-\frac{2y_j(c|c)+y_i(c|c)}{b(c)'}\right\}$$
$$= k\left\{\frac{y(c|c)y_{ji}(c|c)}{b(c)'y_j(c|c)}-\frac{y_i(c|c)}{b(c)'}+\frac{y_j(c|c)}{[b(c)']^2}\right\}$$
$$= \frac{k}{[b(c)']^2}\left\{[b(c)-c]\left[\frac{y_j(c|c)y_i(c|c)}{y(c|c)}-y_{ji}(c|c)\right]+y_j(c|c)\right\}. \qquad (4.38)$$

Because the first derivative of the bid function, see (4.32), and bid shading is positive, see (4.33), the second derivative is negative for all generators if the following condition holds true

$$[b(c)-c]\left[\frac{y_j(c|c)y_i(c|c)}{y(c|c)}-y_{ji}(c|c)\right]+y_j(c|c)<0$$

$$\Longleftrightarrow$$

$$\frac{y_j(c|c)y_i(c|c)}{y(c|c)}-y_{ji}(c|c)<\frac{(-)y_j(c|c)}{b(c)-c}.$$

The inequation, using the property that $y_j(c|c)$ is strictly negative according to (4.25), can be transformed to

$$\begin{aligned}\frac{1}{b(c)-c} &> \frac{y_{ji}(c|c)}{y_j(c|c)} - \frac{y_i(c|c)}{y(c|c)} \\ &> \frac{f_i(c|c)}{f(c|c)} + \frac{[\alpha-\beta]F_i(c|c)}{\alpha-[\alpha-\beta]F(c|c)}.\end{aligned} \tag{4.39}$$

For demand $d \in (0,k]$ with $\beta = 0$ according to (4.2), the condition is given by, taking into account (3.10)

$$\begin{aligned}\frac{1}{b(c)-c} &> \frac{f_i(c|c)}{f(c|c)} + \frac{F_i(c|c)}{1-F(c|c)} \\ &> 0.\end{aligned}$$

The condition is always true because $b(c) > c$ due to (4.33).

Equation (4.39) is for all marginal costs and the demand range $(k, 2k)$ with $\alpha = 1$ and $\beta = \frac{d-k}{k}$, see (4.1) and (4.2), obtained by

$$\frac{1}{b(c)-c} > \frac{f_i(c|c)}{f(c|c)} + \frac{[2k-d]F_i(c|c)}{k-[2k-d]F(c|c)}. \tag{4.40}$$

This expression is valid here due to (4.83).[3] The alternative expression of this condition is presented at the end of this proof, see (4.53).

For independently distributed marginal costs, the condition, using (3.11), (3.12), (3.13), and (3.14), can be rewritten to

$$\begin{aligned}\frac{1}{b^i(c)-c} &> \frac{f_i(c|c)}{f(c|c)} + \frac{[2k-d]F_i(c|c)}{k-[2k-d]F(c|c)} \\ &> 0.\end{aligned}$$

This inequation holds always true because $b^i(c) > c$ due to (4.33).

Step 2: Participation of Generators

An equilibrium only exists if all generators, bidding according to the common bid function, have a weak incentive to participate in the auction. This is given if the following condition holds true

$$\pi(b(c), c) \geq 0. \tag{4.41}$$

[3] Note that this proof also covers the later presented Theorem 4.6, where (4.83) is given. For simplicity, both theorems are proved together due to their same decision problem.

Using the bid function (4.30), the profit function (4.22) is now determined by

$$\pi\left(b\left(c\right),c\right) = k\left[b\left(c\right) - c\right] y\left(c \middle| c\right). \tag{4.42}$$

It is known from (4.33) that the bid function ensures a positive contribution margin. Furthermore, $y\left(c \middle| c\right)$ is positive for all marginal costs except $\bar{c}$, see (4.21). Only a generator with the highest possible marginal costs suffers an expected profit of zero if one bidder is sufficient to serve the entire demand. The only chance to earn money is a standoff. The probability therefore is mathematically zero. Nevertheless, generators do have an incentive to participate in this auction because it was only assumed that they earn non-negative profit, see Assumption 3.2.

Step 3: Existence of a Bayes–Nash Equilibrium

A Bayes–Nash equilibrium exists if no generator is able to improve his profit by making a bid different to that suggested by the bid function while all the others use the common function. The bid function (4.30) maximizes the expected profit under condition (4.39) if no bid lies below the lowest bid or above the highest bid. Hence, profit is not increased if the new bid, which differs from (4.30), is an element of the value range of the bid function: $[b(\underline{c}), b(\bar{c})]$. No deviation will occur inside the range.

Based on this finding, the following relation exists, see (4.42)

$$\begin{aligned} \pi_i\left(b\left(c_i\right),c_i\right) &\geq \pi_i\left(b(\underline{c}),c_i\right) \\ &\geq k\alpha\left[b(\underline{c}) - c_i\right]. \end{aligned} \tag{4.43}$$

The next case of deviation is undercutting slightly the lowest possible bid according to the bid function. This guarantees winning the auction. The expected profit using this strategy is with $\varepsilon \in [b(\underline{c}), 0)$ given by

$$\pi_i\left(b(\underline{c}) - \varepsilon, c_i\right) = k\alpha\left[b(\underline{c}) - \varepsilon - c_i\right]. \tag{4.44}$$

The upper limit of (4.44) is $\pi_i\left(b(\underline{c}), c_i\right)$, see (4.43), because ε was assumed to be greater than zero.

$$\begin{aligned} &\pi_i\left(b(\underline{c}) - \varepsilon, c_i\right) \quad < \pi_i\left(b(\underline{c}), c_i\right) \\ &\Longleftrightarrow \\ &k\alpha\left[b(\underline{c}) - \varepsilon - c_i\right] < k\alpha\left[b(\underline{c}) - c_i\right] \\ &\Longleftrightarrow \\ &0 < \varepsilon. \end{aligned}$$

According to (4.43), this also shows that the profit from undercutting lies below the profit bidding according to the bid function. Therefore, no deviation from the bid function improves the expected profit.

The remaining strategy of deviation is to outbid the highest possible bid. This is only rational if the bid still achieves the conditions of an accepted bid because otherwise the profit is zero, which is worse than using the bid function with a non-negative profit, see (4.41). Therefore, the following must hold true

$$b(\bar{c}) < \bar{p}.$$

Given this, the expected profit of outbidding with $\varepsilon \in (0, \bar{p} - b(\bar{c})]$ is obtained by

$$\pi_i\left(b(\bar{c}) + \varepsilon, c_i\right) = k\beta\left[b(\bar{c}) - c_i + \varepsilon\right] = k\left[b(\bar{c}) - c_i + \varepsilon\right] y\left(\bar{c} \middle| c_i\right). \tag{4.45}$$

It is easy to see, that for $\beta > 0$ at least the generator with the highest marginal costs $\bar{c}$ has an incentive to outbid because he can earn an extra profit, which is $k\beta\left[\bar{p} - b(\bar{c})\right]$ maximal.

This extra profit is lost if the following is taken into account

$$b(\bar{c}) = \bar{p}. \tag{4.46}$$

Due to (4.29), this implies

$$\hat{b} = \bar{p}.$$

In the case of $\beta = 0$ outbidding does not improve the profit as long as $\hat{b}$ is accepted because then the loser produces nothing. The open question is, which $\hat{b}$ should be agreed to. As assumed, generators want to maximize their profits. Therefore, $\hat{b}$ has to be chosen in such a way that it sets the following derivative of the profit function (4.42) to zero

$$\begin{aligned} \frac{\partial \pi_i\left(b\left(c_i\right), c_i\right)}{\partial \hat{b}} &= ky\left(c_i \middle| c_i\right) \frac{\partial b\left(c_i\right)}{\partial \hat{b}} \\ &= ky\left(c_i \middle| c_i\right) \mathrm{e}^{\int_{c_i}^{\bar{c}} \frac{y_j(t|t)}{y(t|t)} dt}. \end{aligned}$$

Obviously, no condition for $\hat{b}$ can be found. According to the profit function (4.42) with respect to bid function (4.30), $\hat{b}$ has to maximize the bid function. The upper limit of the bid function is $\bar{p}$ due to the auction rules. Hence, (4.46) is also valid for $\beta = 0$, which results in $\hat{b} = \bar{p}$. Bid function (4.30) is then determined by

$$b(c) = \left[\bar{p} - \bar{c} + \int_{c}^{\bar{c}} \mathrm{e}^{(-)\int_{\tau}^{\bar{c}} \frac{y_j(t|t)}{y(t|t)} dt} d\tau\right] \mathrm{e}^{\int_{c}^{\bar{c}} \frac{y_j(t|t)}{y(t|t)} dt} + c. \tag{4.47}$$

Step 4: Uniqueness of the Bayes–Nash Equilibrium

The bid function (4.47) contains only variables set by the environment wherein the generator has to decide on his bid. No free variables remain. Hence, only one Bayes–Nash equilibrium exists. Substituting $y(\cdot|\cdot)$ and $y_j(\cdot|\cdot)$, see (4.20) and (4.24), the bid function is obtained by

$$b_d^*(c) = \left\{ \bar{p} - \bar{c} + \int_c^{\bar{c}} e^{\int_\tau^{\bar{c}} \frac{[\alpha-\beta] f(t|t)}{\alpha - [\alpha-\beta] F(t|t)} dt} d\tau \right\} e^{\int_c^{\bar{c}} \frac{(-)[\alpha-\beta] f(t|t)}{\alpha - [\alpha-\beta] F(t|t)} dt} + c. \tag{4.48}$$

If demand lies in the range $(0, k]$ with $\beta = 0$ according to (4.2), the final expression is obtained by

$$b_d^*(c) = \left[\bar{p} - \bar{c} + \int_c^{\bar{c}} e^{\int_\tau^{\bar{c}} \frac{f(t|t)}{1-F(t|t)} dt} d\tau \right] e^{\int_c^{\bar{c}} \frac{(-) f(t|t)}{1-F(t|t)} dt} + c. \tag{4.49}$$

For independently distributed marginal costs, the optimal bid function, using (3.11), (3.12), and (3.3), can be simplified to

$$\begin{aligned} b_d^{*i}(c) &= \left[\bar{p} - \bar{c} + \int_c^{\bar{c}} e^{\int_\tau^{\bar{c}} \frac{f(t)}{1-F(t)} dt} d\tau \right] e^{\int_c^{\bar{c}} \frac{(-) f(t)}{1-F(t)} dt} + c \\ &= \begin{cases} \dfrac{[\bar{p} - \bar{c}][1 - F(\bar{c})] + \int_c^{\bar{c}} [1 - F(\tau)]\, d\tau}{1 - F(c)} + c & : c \in [\underline{c}, \bar{c}) \\ \bar{p} & : c = \bar{c} \end{cases} \\ &= \begin{cases} \dfrac{\bar{c} - c - \int_c^{\bar{c}} F(\tau)\, d\tau}{1 - F(c)} + c & : c \in [\underline{c}, \bar{c}) \\ \bar{p} & : c = \bar{c}. \end{cases} \end{aligned} \tag{4.50}$$

Bid function (4.48) is for demand $d \in (k, 2k)$ with $\alpha = 1$ and $\beta = \frac{d-k}{k}$, see (4.1) and (4.2), given by

$$b_d^*(c) = \left\{ \bar{p} - \bar{c} + \int_c^{\bar{c}} e^{\int_\tau^{\bar{c}} \frac{[2k-d] f(t|t)}{k - [2k-d] F(t|t)} dt} d\tau \right\} e^{\int_c^{\bar{c}} \frac{(-)[2k-d] f(t|t)}{k - [2k-d] F(t|t)} dt} + c. \tag{4.51}$$

Condition (4.40) can now be rewritten to the following expression for all marginal costs, which satisfy

$$0 < \frac{f_i(c|c)}{f(c|c)} + \frac{[2k-d] F_i(c|c)}{k - [2k-d] F(c|c)}, \tag{4.52}$$

because these marginal costs generate positive values at the right term of (4.40), which may cause difficulties. The condition is then obtained by, taking into account (4.33) and using (4.51)

$$\frac{1}{\frac{f_i(c|c)}{f(c|c)}+\frac{[2k-d]F_i(c|c)}{k-[2k-d]F(c|c)}} > b_d^*(c)-c$$

$$> \left\{\bar{p}-\bar{c}+\int_c^{\bar{c}} e^{\int_\tau^{\bar{c}} \frac{[2k-d]f(t|t)}{k-[2k-d]F(t|t)}\,dt}\,d\tau\right\} e^{\int_c^{\bar{c}} \frac{(-)[2k-d]f(t|t)}{k-[2k-d]F(t|t)}\,dt}$$

$$\Longleftrightarrow$$

$$\frac{e^{\int_c^{\bar{c}} \frac{[2k-d]f(t|t)}{k-[2k-d]F(t|t)}\,dt}}{\frac{f_i(c|c)}{f(c|c)}+\frac{[2k-d]F_i(c|c)}{k-[2k-d]F(c|c)}} > \bar{p}-\bar{c}+\int_c^{\bar{c}} e^{\int_\tau^{\bar{c}} \frac{[2k-d]f(t|t)}{k-[2k-d]F(t|t)}\,dt}\,d\tau$$

$$\Longleftrightarrow$$

$$\bar{p} < \bar{c}+\min\left(e^{\int_c^{\bar{c}} \frac{[2k-d]f(t|t)}{k-[2k-d]F(t|t)}\,dt}\left\{\frac{1}{\frac{f_i(c|c)}{f(c|c)}+\frac{[2k-d]F_i(c|c)}{k-[2k-d]F(c|c)}}\right.\right.$$
$$\left.\left.-\int_c^{\bar{c}} e^{\int_c^{\tau} \frac{(-)[2k-d]f(t|t)}{k-[2k-d]F(t|t)}\,dt}\,d\tau\right\}\right). \tag{4.53}$$

A simpler expression of the optimal bid, using (3.11), (3.12), and (3.3), is for independently distributed marginal costs obtained by

$$b_d^{*i}(c) = \left\{\bar{p}-\bar{c}+\int_c^{\bar{c}} e^{\int_\tau^{\bar{c}} \frac{[2k-d]f(t)}{k-[2k-d]F(t)}\,dt}\,d\tau\right\} e^{\int_c^{\bar{c}} \frac{(-)[2k-d]f(t)}{k-[2k-d]F(t)}\,dt}+c$$

$$= [\bar{p}-\bar{c}]\frac{k-[2k-d]F(\bar{c})}{k-[2k-d]F(c)}+\frac{\int_c^{\bar{c}}\{k-[2k-d]F(\tau)\}\,d\tau}{k-[2k-d]F(c)}+c$$

$$= \frac{[d-k][\bar{p}-\bar{c}]+k[\bar{c}-c]-[2k-d]\int_c^{\bar{c}} F(\tau)\,d\tau}{k-[2k-d]F(c)}+c$$

$$= \frac{[d-k][\bar{p}-c]+[2k-d]\int_c^{\bar{c}}[1-F(\tau)]\,d\tau}{k-[2k-d]F(c)}+c. \tag{4.54}$$

□

Interesting properties of the optimal bid function are summarized in Corollary 4.1.

Corollary 4.1. *Suppose $d \in (0,k]$.*

(a) *Bid shading is positive, i.e. the optimal bid of a generator is greater than his marginal costs.*
(b) *The optimal bid function increases strictly with marginal costs.*
(c) *Optimal bids are independent of demand.*

Proof.

(a) Optimal bids are greater than marginal costs as (4.33) states. Theorem 4.2 proved that this condition holds true. Hence, optimal bids do not represent the true marginal costs. Additionally, bid shading, i.e. the difference of bid minus marginal costs, is positive.
(b) A condition of the proof of Theorem 4.2 was that the optimal bid function must be strictly increasing with respect to marginal costs. It was shown there that this condition holds true, see (4.32).
(c) The first derivative of the bid function with respect to demand, using (4.18), is given by

$$\frac{\partial b_d^*(c)}{\partial d} = 0. \tag{4.55}$$

Hence, the optimal bid function does not change with demand. □

This shows that generators do not bid their true marginal costs, they bid higher, see Corollary 4.1 (a). This ensures that generators stay in the market because they get paid their own bids according to the auction rules. Lower bids would lead to zero profits or even losses. The consequence would be a market failure because of a lack of generators. The least productive generator with the highest possible marginal cost $\bar{c}$ expects only a profit of zero. This may motivates him to leave the market. But from the economic point of view, he has after all one chance of realizing a positive profit, the standoff.

The bid function increases with marginal costs according to Corollary 4.1 (b). This covers the fact that low production costs are a competitive advantage, which leads to lower bids with a greater chance of winning. That bids are independent of demand, see Corollary 4.1 (c), is a result of the strong competitive pressure because the winner takes it all and the loser gets nothing. This is also a consequence of demand-independent variable costs.

Example 4.2.

Suppose the probability distribution given by Definition 3.5. The optimal bid function for independently distributed marginal costs is given by

$$b_d^{*i}(c) = \begin{cases} \frac{1+c}{2} & : c \in [0,1) \\ \bar{p} & : c = 1. \end{cases} \tag{4.56}$$

For strong affiliated marginal costs with $a > 0$, the bid function is obtained by

$$b_d^*(c) = \left\{ \bar{p} - 1 + \int_c^1 \exp\left(\int_\tau^1 \frac{2\left[1 + at^2\right]}{2 + at - [2 + at^2]\,t}\, dt \right) d\tau \right\}$$

$$\times \exp\left(- \int_c^1 \frac{2\left[1 + at^2\right]}{2 + at - [2 + at^2]\,t}\, dt \right) + c. \tag{4.57}$$

Proof.

The optimal bid function (4.18), taking into account (3.26) and (3.27), is obtained by

$$b_d^*(c) = \left[\bar{p} - 1 + \int_c^1 \mathrm{e}^{\int_\tau^1 \frac{\frac{1+at^2}{1+\frac{a}{2}t}}{1 - \frac{1+\frac{a}{2}t^2}{1+\frac{a}{2}t}\, t} dt} d\tau \right] \mathrm{e}^{(-)\int_c^1 \frac{\frac{1+at^2}{1+\frac{a}{2}t}}{1 - \frac{1+\frac{a}{2}t^2}{1+\frac{a}{2}t}\, t} dt} + c$$

$$= \left\{ \bar{p} - 1 + \int_c^1 \mathrm{e}^{\int_\tau^1 \frac{2\left[1+at^2\right]}{2+at-\left[2+at^2\right]t} dt} d\tau \right\} \mathrm{e}^{\int_c^1 \frac{(-2)\left[1+at^2\right]}{2+at-\left[2+at^2\right]t} dt} + c. \qquad \square$$

Figure 4.2 shows the bid functions for independently distributed marginal costs in the first graph and for strongly affiliated costs with $a = 4$ in the second graph. Bids are lower for independently distributed costs. This is reasonable because the probability of having the lowest marginal costs increases with the degree of affiliation, see the plot of (3.33) in Figure 3.3. The higher probability of winning increases the expected amount of output sold in the case of winning, which is only important here because the loser sells nothing. It allows a generator to raise his bid because his competitor also has now more to lose. Hence, stronger affiliation reduces competitive pressure expressed in higher bids.

So far we know the generators' bidding behavior. But does this induce an efficient input factor allocation by using the auction procedure? From the economic point of view, it is desirable that minimal production costs always result. Theorem 4.3 answers such questions relying on the general results of Theorem 4.1.

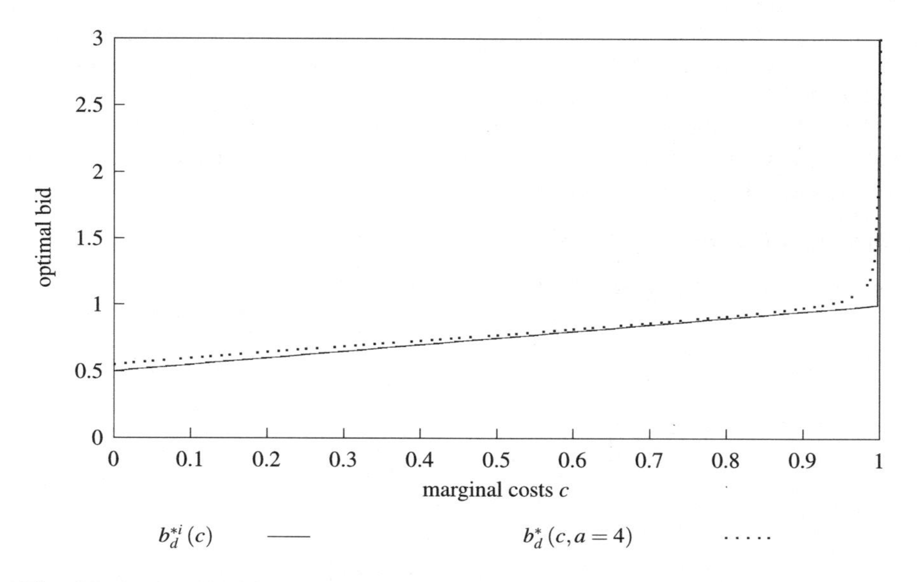

Fig. 4.2 Optimal bid function in a discriminatory auction (case A)

Theorem 4.3. *Suppose* $d \in (0,k]$.

(a) *The auction leads to production efficiency. Production costs are the lowest possible.*

(b) *The expected production costs per electricity unit are given by*

$$E[c_d] = 2\int_{\underline{c}}^{\bar{c}} cF(c)\left[f(c|c) + F_i(c|c)\right]dc$$
$$-2\int_{\underline{c}}^{\bar{c}} F(c)\left[1 - F(c|c)\right]dc. \tag{4.58}$$

For independently distributed marginal costs, the expression reduces to

$$E\left[c_d^i\right] = \bar{c} + \int_{\underline{c}}^{\bar{c}} F(c)^2\,dc - 2\int_{\underline{c}}^{\bar{c}} F(c)\,dc. \tag{4.59}$$

Proof.

(a) Optimal bid function (4.18) increases with marginal costs, see (4.32). Resulting bids are accepted bids according to the auction rules. Hence, Theorem 4.1 applies, especially (a).

(b) Equations (4.4) and (4.5) can be used because the same reason that is given in (a) applies. □

Generators are scheduled in such a way that production efficiency is achieved. This is the first statement of Theorem 4.3. The theorem also presents expressions for the calculation of expected production costs.

The following theorem contains formulas for the expected profit and contribution margin. They are of interest because generators can use them to compare different auction types and make a preference list.

Theorem 4.4. *Suppose* $d \in (0,k]$.

(a) *The expected profit of a generator is obtained by*

$$\pi_d^*(c) = d[1-F(c|c)]\left[\bar{p}-\bar{c}+\int_c^{\bar{c}} \exp\left(\int_\tau^{\bar{c}} \frac{f(t|t)}{1-F(t|t)}\,dt\right) d\tau\right] \times \exp\left(-\int_c^{\bar{c}} \frac{f(t|t)}{1-F(t|t)}\,dt\right). \tag{4.60}$$

The expression for independently distributed marginal costs reduces to

$$\pi_d^{*i}(c) = d\left[\bar{c}-c-\int_c^{\bar{c}} F(\tau)\,d\tau\right]. \tag{4.61}$$

(b) *The expected contribution margin per capacity unit is obtained by*

$$\pi_d^{u^*}(c) = \frac{d}{k}[1-F(c|c)]\left[\bar{p}-\bar{c}+\int_c^{\bar{c}} \exp\left(\int_\tau^{\bar{c}} \frac{f(t|t)}{1-F(t|t)}\,dt\right) d\tau\right] \times \exp\left(-\int_c^{\bar{c}} \frac{f(t|t)}{1-F(t|t)}\,dt\right). \tag{4.62}$$

If marginal costs are independently drawn, the expression reduces to

$$\pi_d^{u^{*i}}(c) = \frac{d}{k}\left[\bar{c}-c-\int_c^{\bar{c}} F(\tau)\,d\tau\right]. \tag{4.63}$$

Proof.

(a) Optimal bid function (4.18) is used to derive the expected profit, see (4.22). The expected profit is obtained by, taking into account (4.20), (4.1), and (4.3)

$$\pi_d^*(c) = k[b(c)-c]\,y(c|c)$$

$$= k\left[\bar{p}-\bar{c}+\int_{c}^{\bar{c}} \mathrm{e}^{\int_{\tau}^{\bar{c}} \frac{f(t|t)}{1-F(t|t)}\,dt}\,d\tau\right] \mathrm{e}^{\int_{c}^{\bar{c}} \frac{(-)f(t|t)}{1-F(t|t)}\,dt}\, y(c|c)$$

$$= k\left[\alpha-\beta_f F(c|c)\right]\left[\bar{p}-\bar{c}+\int_{c}^{\bar{c}} \mathrm{e}^{\int_{\tau}^{\bar{c}} \frac{f(t|t)}{1-F(t|t)}\,dt}\,d\tau\right] \mathrm{e}^{\int_{c}^{\bar{c}} \frac{(-)f(t|t)}{1-F(t|t)}\,dt}$$

$$= d[1-F(c|c)]\left[\bar{p}-\bar{c}+\int_{c}^{\bar{c}} \mathrm{e}^{\int_{\tau}^{\bar{c}} \frac{f(t|t)}{1-F(t|t)}\,dt}\,d\tau\right] \mathrm{e}^{\int_{c}^{\bar{c}} \frac{(-)f(t|t)}{1-F(t|t)}\,dt}.$$

The profit for independently distributed marginal costs, using (3.11), (3.12), and (3.3), is given by

$$\pi_d^{*i}(c) = d[1-F(c)]\left[\bar{p}-\bar{c}+\int_{c}^{\bar{c}} \mathrm{e}^{\int_{\tau}^{\bar{c}} \frac{f(t)}{1-F(t)}\,dt}\,d\tau\right] \mathrm{e}^{\int_{c}^{\bar{c}} \frac{(-)f(t)}{1-F(t)}\,dt}$$

$$= d\int_{c}^{\bar{c}} [1-F(\tau)]\,d\tau$$

$$= d\left[\bar{c}-c-\int_{c}^{\bar{c}} F(\tau)\,d\tau\right].$$

(b) The expected contribution margin per electricity unit, using (4.60), is given by

$$\pi_d^{u^*}(c) = \frac{\pi_d^*(c)}{k}$$

$$= \frac{d}{k}[1-F(c|c)]\left[\bar{p}-\bar{c}+\int_{c}^{\bar{c}} \mathrm{e}^{\int_{\tau}^{\bar{c}} \frac{f(t|t)}{1-F(t|t)}\,dt}\,d\tau\right] \mathrm{e}^{\int_{c}^{\bar{c}} \frac{(-)f(t|t)}{1-F(t|t)}\,dt}.$$

It is easy to see that for independently distributed costs, using (4.61), the expression is simply

$$\pi_d^{u^{*i}}(c) = \frac{d}{k}\left[\bar{c}-c-\int_{c}^{\bar{c}} F(\tau)\,d\tau\right].$$

□

Corollary 4.2. *Suppose $d \in (0,k]$.*

(a) *The expected profit and expected contribution margin per capacity unit are positive for $c \in [\underline{c},\bar{c})$ and zero for the highest marginal cost $\bar{c}$.*

(b) *The expected profit and expected contribution margin increase strictly with demand for $c \in [\underline{c},\bar{c})$. Both expressions are independent of demand for $\bar{c}$.*

(c) *The expected profit and expected contribution margin strictly decrease with marginal costs for independent distribution. Otherwise, the same statement hold true if the following condition is satisfied for all* $c \in (\underline{c}, \bar{c})$

$$\frac{1}{b_d^*(c) - c} > \frac{(-)F_i(c|c)}{1 - F(c|c)}. \tag{4.64}$$

The slope of both expressions is zero at $\bar{c}$ *even for independent distribution.*

Proof.

(a) The expressions of the expected profit, see (4.60), and the expected contribution margin, see (4.62), are zero for $\bar{c}$ because the probability of winning $1 - F(\bar{c}|\bar{c})$ is zero due to (3.3). This probability is positive for all other marginal costs. The factor

$$\left[\bar{p} - \bar{c} + \int_c^{\bar{c}} e^{\int_\tau^{\bar{c}} \frac{f(t|t)}{1-F(t|t)} dt} d\tau\right] e^{\int_c^{\bar{c}} \frac{(-)f(t|t)}{1-F(t|t)} dt},$$

which is found in both expressions, represents the bid shading or the difference between optimal bid minus the marginal costs. It is positive due to Corollary 4.1 (a). Hence, the expected profit and expected contribution margin are positive for $c \in [\underline{c}, \bar{c})$ because additionally d and k are greater than zero.

(b) The first derivative of the expected profit with respect to demand, using (4.60), is given by

$$\frac{\partial \pi_d^*(c)}{\partial d} = [1 - F(c|c)] \left[\bar{p} - \bar{c} + \int_c^{\bar{c}} e^{\int_\tau^{\bar{c}} \frac{f(t|t)}{1-F(t|t)} dt} d\tau\right] e^{\int_c^{\bar{c}} \frac{(-)f(t|t)}{1-F(t|t)} dt}. \tag{4.65}$$

A simpler expression for the proof, using (4.18), is obtained by

$$\frac{\partial \pi_d^*(c)}{\partial d} = [1 - F(c|c)][b_d^*(c) - c].$$

The second factor is always positive due to Corollary 4.1 (a). The first factor is positive for $c \in [\underline{c}, \bar{c})$ and zero for $\bar{c}$ because of (3.3). Hence, the first derivative is positive and constant for $c \in [\underline{c}, \bar{c})$. The expected profit is independent of demand for $\bar{c}$ because the first derivative is zero due to (3.3).

The slope for the expected contribution margin, using (4.62), is obtained by

$$\frac{\partial \pi_d^{u^*}(c)}{\partial d} = \frac{1}{k} \frac{\partial \pi_d^*(c)}{\partial d}. \tag{4.66}$$

The derivative is also positive because of the same argument given for the expected profit.

(c) The expected profit decreases strictly with marginal costs if the first derivative with respect to marginal costs is negative for $c \in [\underline{c}, \bar{c})$ and zero for $\bar{c}$. Using (4.60) and (4.18), the first derivative is given by

$$\frac{\partial \pi_d^*(c)}{\partial c} = (-)d[f(c|c) + F_i(c|c)][b_d^*(c) - c] + d[1 - F(c|c)][b_d^*(c)' - 1].$$

Using (4.31), (4.24), (4.20), and (4.1)–(4.3), the term is rewritten to

$$\begin{aligned} \frac{\partial \pi_d^*(c)}{\partial c} &= (-)d[f(c|c) + F_i(c|c)][b_d^*(c) - c] \\ &\quad + d[1 - F(c|c)] \left\{ [b_d^*(c) - c] \frac{f(c|c)}{1 - F(c|c)} - 1 \right\} \\ &= (-)d\{[b_d^*(c) - c]F_i(c|c) + 1 - F(c|c)\}. \end{aligned} \tag{4.67}$$

The slope of the expected profit is zero at $\bar{c}$ due to (3.5) and (3.8). Hence, profit decreases strictly with marginal costs if the following holds true for $c \in [\underline{c}, \bar{c})$

$$\begin{aligned} &\frac{\partial \pi_d^*(c)}{\partial c} && < 0 \\ &\Longleftrightarrow \\ &0 && < [b_d^*(c) - c]F_i(c|c) + 1 - F(c|c) \\ &\Longleftrightarrow \\ &\frac{1}{b_d^*(c) - c} && > \frac{(-)F_i(c|c)}{1 - F(c|c)}. \end{aligned} \tag{4.68}$$

The condition is for independently distributed marginal costs given by, taking into account $F_i(c|c) = 0$, see (3.14)

$$\frac{1}{b_d^*(c) - c} > 0.$$

This is valid because bids are greater than marginal costs, see Corollary 4.1 (a). Hence, the expected profit strictly decreases with marginal costs for the independent distribution, while the slope at $\bar{c}$ is zero.

Similar arguments are found for the expected contribution margin because factor $\frac{1}{k}$ is independent of c. □

The generator with the highest marginal costs expects to earn nothing, see Corollary 4.2 (a), because his probability to win $1 - F(\bar{c}|\bar{c})$ is zero. The only chance to earn something is a standoff. It is not mathematically relevant although the realized profit in this case is positive because the bid is greater than marginal costs, see Corollary 4.1 (a). The expected zero profit is the reason why the profit is independent of demand for $\bar{c}$. All other generators with lower marginal costs face an increasing expected profit with higher demand, see Corollary 4.2 (b), due to more electricity sold for the same marginal costs. The better expected contribution margin is the result of a higher utilization of production capacity.

Example 4.3. *Suppose the probability distribution given by Definition 3.5.*

(a) *The expected profit of a generator for independently distributed marginal costs is given by*

$$\pi_d^{*i}(c) = d\,\frac{[1-c]^2}{2}. \tag{4.69}$$

The same expression for strong affiliation with $a > 0$ is obtained by

$$\pi_d^{*}(c) = d\left[1 - \frac{2+ac^2}{2+ac}c\right]\exp\left(-\int_c^1 \frac{2\left[1+at^2\right]}{2+at-[2+at^2]t}dt\right)$$
$$\times\left\{\bar{p} - 1 + \int_c^1 \exp\left(\int_\tau^1 \frac{2\left[1+at^2\right]}{2+at-[2+at^2]t}dt\right)d\tau\right\}. \tag{4.70}$$

(b) *With independently distributed marginal costs, the expected contribution margin per capacity unit is obtained by*

$$\pi_d^{u*i}(c) = \frac{d}{k}\,\frac{[1-c]^2}{2}. \tag{4.71}$$

The contribution margin for strong affiliation with $a > 0$ is given by

$$\pi_d^{u*}(c) = \frac{d}{k}\left[1 - \frac{2+ac^2}{2+ac}c\right]\exp\left(-\int_c^1 \frac{2\left[1+at^2\right]}{2+at-[2+at^2]t}dt\right)$$
$$\times\left\{\bar{p} - 1 + \int_c^1 \exp\left(\int_\tau^1 \frac{2\left[1+at^2\right]}{2+at-[2+at^2]t}dt\right)d\tau\right\}. \tag{4.72}$$

Proof.

(a) The expected profit of a generator (4.60), using (3.26) and (3.27), is given by

$$\pi_d^*(c) = d\left[1 - \frac{1+\frac{a}{2}c^2}{1+\frac{a}{2}c}c\right]\left\{\bar{p} - 1 + \int_c^1 e^{\int_\tau^1 \frac{\frac{1+at^2}{1+\frac{a}{2}t}}{1-\frac{1+\frac{a}{2}t^2}{1+\frac{a}{2}t}t}dt} d\tau\right\}$$

$$\times e^{(-)\int_c^1 \frac{\frac{1+at^2}{1+\frac{a}{2}t}}{1-\frac{1+\frac{a}{2}t^2}{1+\frac{a}{2}t}t}dt}$$

$$= d\left[1 - \frac{2+ac^2}{2+ac}c\right]\left\{\bar{p} - 1 + \int_c^1 e^{\int_\tau^1 \frac{2[1+at^2]}{2+at-[2+at^2]t}dt} d\tau\right\}$$

$$\times e^{\int_c^1 \frac{(-2)[1+at^2]}{2+at-[2+at^2]t}dt}.$$

The expression for independently distributed marginal costs, using (4.61) and (3.25), is obtained by

$$\pi_d^{*i}(c) = d\left[1 - c - \int_c^1 \tau\, d\tau\right]$$

$$= d\left[1 - c - \frac{1-c^2}{2}\right] = d\,\frac{[1-c]^2}{2}.$$

(b) The expected contribution margin per capacity unit (4.62), using (4.70), is given by

$$\pi_d^{u^*}(c) = \frac{\pi_d^*(c)}{k}$$

$$= \frac{d}{k}\left[1 - \frac{2+ac^2}{2+ac}c\right]\left\{\bar{p} - 1 + \int_c^1 e^{\int_\tau^1 \frac{2[1+at^2]}{2+at-[2+at^2]t}dt} d\tau\right\}$$

$$\times e^{\int_c^1 \frac{(-2)[1+at^2]}{2+at-[2+at^2]t}dt}.$$

For independently distributed marginal costs, the expression, using (4.69), is simply

$$\pi_d^{u^{*i}}(c) = \frac{\pi_d^{*i}(c)}{k}$$

$$= \frac{d}{k}\,\frac{[1-c]^2}{2}.$$

□

The expected contribution margins for independently drawn marginal costs and for strong affiliated costs with $a = 4$ are shown in Figure 4.3. The lower two graphs represent a demand of 0.4 of the production capacity of a single generator. The remaining two graphs show results for a full utilization of the winner.

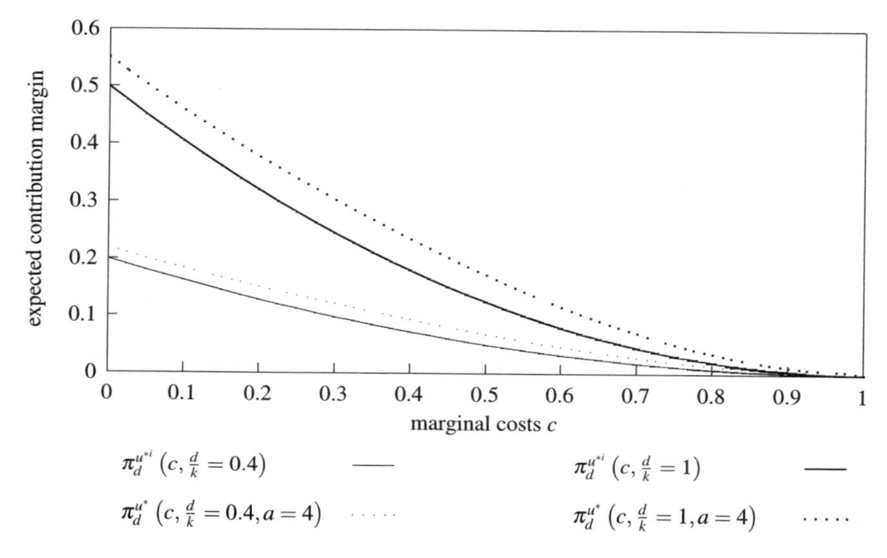

Fig. 4.3 Expected contribution margin in a discriminatory auction (case A)

The expected contribution margin decreases in the example with higher marginal costs, independently of demand. It shows that low marginal costs is a competitive advantage. It pays off resulting in higher expected profits.

The expected contribution margins for strong affiliation dominate those for independently distributed costs. The main reason is that generators bid higher with strong affiliated marginal costs as Figure 4.2 shows. They can do so because they know better the competitive pressure by looking at their own marginal costs. More effectively used information, which is free because the distribution is known, lead to better bidding strategies with higher bids.

Consumers are interested in low prices. The expected price per electricity unit represents the cost they face on average over all possible realizations of marginal costs. Because marginal costs are the generators' confidential information, risk-neutral consumers should compare expected prices if they can decide on the auction type that should be implemented in the market.

Theorem 4.5. *Suppose $d \in (0,k]$. The expected average price is given by*

$$E[p_d] = 2\int_{\underline{c}}^{\bar{c}} cF(c)\,[f(c|c)+F_i(c|c)]\,dc + 2\int_{\underline{c}}^{\bar{c}} F(c)\,F_i(c|c)$$
$$\times \left[\bar{p}-\bar{c}+\int_{c}^{\bar{c}} \exp\left(\int_{\tau}^{\bar{c}} \frac{f(t|t)}{1-F(t|t)}\,dt\right)d\tau\right]$$
$$\times \exp\left(-\int_{c}^{\bar{c}} \frac{f(t|t)}{1-F(t|t)}\,dt\right)dc. \tag{4.73}$$

For independently distributed marginal costs, the expression reduces to

$$E\left[p_d^i\right] = \bar{c} - \int_{\underline{c}}^{\bar{c}} F(c)^2\,dc. \tag{4.74}$$

Proof.

Similar to the proof of Theorem 4.2, this proof is done for the wider demand range $d \in (0,2k)$. The expected average price paid for a unit of electricity is, with $j \neq i$, given by

$$E[p_d] = \frac{1}{\alpha+\beta}E\left[\alpha b_d^*(c_i)\,1_{\{b_d^*(c_i)<b_d^*(c_j)\}} + \beta b_d^*(c_i)\,1_{\{b_d^*(c_i)>b_d^*(c_j)\}}\right]$$
$$+\frac{1}{\alpha+\beta}E\left[\alpha b_d^*(c_j)\,1_{\{b_d^*(c_j)<b_d^*(c_i)\}} + \beta b_d^*(c_j)\,1_{\{b_d^*(c_j)>b_d^*(c_i)\}}\right].$$

Due to identical bid functions and because information variables are drawn from the same distribution, the expected average price is obtained by

$$E[p_d] = \frac{2\alpha}{\alpha+\beta}E\left[b_d^*(c_i)\,1_{\{b_d^*(c_i)<b_d^*(c_j)\}}\right]$$
$$+\frac{2\beta}{\alpha+\beta}E\left[b_d^*(c_i)\,1_{\{b_d^*(c_i)>b_d^*(c_j)\}}\right]$$
$$= \frac{2\alpha}{\alpha+\beta}E\left[b_d^*(c_i)\right] - \frac{\alpha-\beta}{\alpha+\beta}E\left[2b_d^*(c_i)\,1_{\{b_d^*(c_i)>b_d^*(c_j)\}}\right].$$

This is simplified to following expression, because the bid function is strictly increasing in c, see (4.32). The first derivative of the bid function (4.31) is used as well as (4.24), (4.34), and (4.3).

$$
\begin{aligned}
E[p_d] &= \frac{2\alpha}{\alpha+\beta}\int_{\underline{c}}^{\bar{c}} b_d^*(c)\int_{\underline{c}}^{\bar{c}} f(\xi,c)\,d\xi\,dc - \frac{2\beta_f}{\alpha+\beta}\int_{\underline{c}}^{\bar{c}} b_d^*(c)\int_{\underline{c}}^{c} f(\xi,c)\,d\xi\,dc \\
&= \frac{2}{\alpha+\beta}\int_{\underline{c}}^{\bar{c}} b_d^*(c)\,f(c)\,y(c|c)\,dc \qquad (4.75)\\
&= \frac{2}{\alpha+\beta}\,b_d^*(c)\,F(c)\,y(c|c)\Big|_{\underline{c}}^{\bar{c}} - \frac{2}{\alpha+\beta}\int_{\underline{c}}^{\bar{c}} b_d^*(c)'\,F(c)\,y(c|c)\,dc \\
&\quad - \frac{2}{\alpha+\beta}\int_{\underline{c}}^{\bar{c}} b_d^*(c)\,F(c)\,[y_j(c|c)+y_i(c|c)]\,dc \\
&= \frac{2\beta}{\alpha+\beta}\,\bar{p} - \frac{2}{\alpha+\beta}\int_{\underline{c}}^{\bar{c}} b_d^*(c)\,F(c)\,[y_j(c|c)+y_i(c|c)]\,dc \\
&\quad + \frac{2}{\alpha+\beta}\int_{\underline{c}}^{\bar{c}} [b_d^*(c)-c]\,F(c)\,y_j(c|c)\,dc \\
&= \frac{2\beta}{\alpha+\beta}\,\bar{p} - \frac{2}{\alpha+\beta}\int_{\underline{c}}^{\bar{c}} F(c)\left[c y_j(c|c) + b_d^*(c)\,y_i(c|c)\right] dc \\
&= \frac{2\beta}{\alpha+\beta}\,\bar{p} + \frac{2\beta_f}{\alpha+\beta}\int_{\underline{c}}^{\bar{c}} F(c)\left[c f(c|c) + b_d^*(c)\,F_i(c|c)\right] dc.
\end{aligned}
$$

If demand lies in the range $(0,k]$ with $\beta = 0$ and $\beta_f = \alpha$, see (4.2) and (4.3), the expected average price, using bid function (4.18), is obtained by

$$
\begin{aligned}
E[p_d] &= 2\int_{\underline{c}}^{\bar{c}} F(c)\left[c f(c|c) + b_d^*(c)\,F_i(c|c)\right] dc \\
&= 2\int_{\underline{c}}^{\bar{c}} F(c)\,F_i(c|c)\left\{\left[\bar{p}-\bar{c}+\int_{c}^{\bar{c}} e^{\int_{\tau}^{\bar{c}} \frac{f(t|t)}{1-F(t|t)}\,dt}\,d\tau\right] e^{\int_{c}^{\bar{c}} \frac{(-)f(t|t)}{1-F(t|t)}\,dt} + c\right\} dc \\
&\quad +2\int_{\underline{c}}^{\bar{c}} cF(c)\,f(c|c)\,dc \\
&= 2\int_{\underline{c}}^{\bar{c}} F(c)\,F_i(c|c)\left[\bar{p}-\bar{c}+\int_{c}^{\bar{c}} e^{\int_{\tau}^{\bar{c}} \frac{f(t|t)}{1-F(t|t)}\,dt}\,d\tau\right] e^{\int_{c}^{\bar{c}} \frac{(-)f(t|t)}{1-F(t|t)}\,dt}\,dc \\
&\quad +2\int_{\underline{c}}^{\bar{c}} cF(c)\,[F_i(c|c)+f(c|c)]\,dc.
\end{aligned}
$$

The expression for independently distributed marginal costs, see (3.11), (3.14), and (3.3), is simply

$$
\begin{aligned}
E\left[p_d^i\right] &= 2\int_{\underline{c}}^{\bar{c}} cF(c)\,f(c)\,dc = cF(c)^2\Big|_{\underline{c}}^{\bar{c}} - \int_{\underline{c}}^{\bar{c}} F(c)^2\,dc \\
&= \bar{c} - \int_{\underline{c}}^{\bar{c}} F(c)^2\,dc.
\end{aligned}
$$

The expected average price for $d \in (k, 2k)$, using (4.1), (4.2), (4.3), and bid function (4.51), is given by

$$
\begin{aligned}
E\left[p_d\right] &= 2\left[1-\frac{k}{d}\right]\bar{p} + 2\left[\frac{2k}{d}-1\right]\int_{\underline{c}}^{\bar{c}} F(c)\left[cf(c|c) + b_d^*(c)\,F_i(c|c)\right]dc \\
&= 2\left[1-\frac{k}{d}\right]\bar{p} + 2\left[\frac{2k}{d}-1\right]\left(\int_{\underline{c}}^{\bar{c}} cF(c)\left[f(c|c) + F_i(c|c)\right]dc\right. \\
&\quad + \int_{\underline{c}}^{\bar{c}} F(c)\,F_i(c|c)\left\{\bar{p} - \bar{c} + \int_{c}^{\bar{c}} e^{\int_{\tau}^{\bar{c}} \frac{[2k-d]f(t|t)}{k-[2k-d]F(t|t)}\,dt}\,d\tau\right\} \\
&\quad \left.\times e^{\int_{c}^{\bar{c}} \frac{(-)[2k-d]f(t|t)}{k-[2k-d]F(t|t)}\,dt}\,dc\right).
\end{aligned}
\tag{4.76}
$$

The average price for independently distributed costs, see (3.11), (3.14), (3.3), (4.77), as well as bid function (4.54), is obtained by

$$
\begin{aligned}
E\left[p_d^i\right] &= 2\left[1-\frac{k}{d}\right]\bar{p} + 2\left[\frac{2k}{d}-1\right]\int_{\underline{c}}^{\bar{c}} cF(c)\,f(c)\,dc \\
&= 2\left[1-\frac{k}{d}\right]\bar{p} + \left[\frac{2k}{d}-1\right]\left[\bar{c} - \int_{\underline{c}}^{\bar{c}} F(c)^2\,dc\right].
\end{aligned}
\tag{4.77}
$$

□

Corollary 4.3. *Suppose $d \in (k, 2k)$.*

(a) *The expected average price is greater than the expected production costs.*
(b) *The expected average price is independent of demand.*

Proof.

(a) The expected average price is greater than the expected production costs if the following holds true, taking into account that the expected production costs (4.58) equals (4.4)

$$0 < E[p_d] - E[c].$$

The expected average price (4.73), using (4.75), (4.1)–(4.3), and (4.20), is rewritten to

$$E[p_d] = 2\int_{\underline{c}}^{\bar{c}} b_d^*(c)\, f(c)\,[1 - F(c|\,c)]\, dc.$$

The lowest expected cost (4.4), using (4.9), (4.1), and (4.2), is given by

$$E[c] = 2\int_{\underline{c}}^{\bar{c}} c f(c)\,[1 - F(c|\,c)]\, dc. \tag{4.78}$$

The average price is therefore greater than production costs if

$$0 < 2\int_{\underline{c}}^{\bar{c}} [b_d^*(c) - c] f(c)\,[1 - F(c|\,c)]\, dc$$

is valid. It is sufficient that following holds true, taking into account (3.5) and Assumption 3.7, declaring that the probability density function is positive for all marginal costs

$$b_d^*(c) > c.$$

This relation was proved by Corollary 4.1 (a). Therefore, the expected average price is greater than the expected production costs.

(b) The first derivative of the expected average price is obtained by, using (4.110)

$$\frac{\partial E[p_d]}{\partial d} = 0. \tag{4.79}$$

Hence, the price is independent of demand. □

The expected average price must be greater than the marginal costs, see Corollary 4.3 (a), because otherwise no generator would earn money and would leave the market. This would result in a market failure. It is not surprising that the average price is independent of demand because it is determined only by the winning bids. Optimal bids are demand-independent due to Corollary 4.1 (c). This property is preserved because the probability, which is used as the weight for the average price, is also independent of demand.

Example 4.4.

Suppose the probability distribution given by Definition 3.5. The expected average price for independently distributed marginal costs is given by

$$E\left[p_d^i\right] = \frac{2}{3}. \tag{4.80}$$

For strong affiliation with $a > 0$, the expected average price is obtained by

$$\begin{aligned} E\left[p_d\right] &= \frac{4}{5} - \frac{2}{4+a}\left[\frac{1}{10} + \frac{2}{3a} - \frac{4}{a^2} - \frac{16}{a^3} + 16\frac{2+a}{a^4}\ln\left(1+\frac{a}{2}\right)\right] \\ &\quad - \frac{4a}{4+a}\int_0^1 \frac{4+ac}{[2+ac]^2}[1-c]c^2 \exp\left(-\int_c^1 \frac{2\left[1+at^2\right]}{2+at-[2+at^2]t}dt\right) \\ &\quad \times \left\{\bar{p} - 1 + \int_c^1 \exp\left(\int_\tau^1 \frac{2\left[1+at^2\right]}{2+at-[2+at^2]t}dt\right)d\tau\right\}dc. \end{aligned} \tag{4.81}$$

Proof.

The expected average price for independently distributed marginal costs, using (4.74) and (3.25), is obtained by

$$E\left[p_d^i\right] = 1 - \int_0^1 c^2\,dc = 1 - \frac{1}{3}c^3\Big|_0^1 = \frac{2}{3}.$$

For strong affiliated costs with $a > 0$, the average price, using (4.73), (4.17), (3.25), (3.27), and bid function (4.57), is given by

$$\begin{aligned} E\left[p_d\right] &= \frac{4}{5} - \frac{2}{4+a}\left[\frac{1}{10} + \frac{2}{3a} - \frac{4}{a^2} - \frac{16}{a^3} + 16\frac{2+a}{a^4}\ln\left(1+\frac{a}{2}\right)\right] \\ &\quad - \frac{a}{2}\int_0^1 \frac{1+\frac{a}{4}c}{1+\frac{a}{4}}\frac{1-c}{\left[1+\frac{a}{2}c\right]^2}c^2\left\{\bar{p} - 1 + \int_c^1 e^{\int_\tau^1 \frac{2\left[1+at^2\right]}{2+at-\left[2+at^2\right]t}dt}d\tau\right\} \\ &\quad \times e^{\int_c^1 \frac{(-2)\left[1+at^2\right]}{2+at-\left[2+at^2\right]t}dt}dc \\ &= \frac{4}{5} - \frac{2}{4+a}\left[\frac{1}{10} + \frac{2}{3a} - \frac{4}{a^2} - \frac{16}{a^3} + 16\frac{2+a}{a^4}\ln\left(1+\frac{a}{2}\right)\right] \\ &\quad - \frac{4a}{4+a}\int_0^1 \frac{4+ac}{[2+ac]^2}[1-c]c^2\left\{\bar{p} - 1 + \int_c^1 e^{\int_\tau^1 \frac{2\left[1+at^2\right]}{2+at-\left[2+at^2\right]t}dt}d\tau\right\} \\ &\quad \times e^{\int_c^1 \frac{(-2)\left[1+at^2\right]}{2+at-\left[2+at^2\right]t}dt}dc. \end{aligned}$$

□

Because demand does not have an impact on the expected price or production costs, the degree of affiliation (a) is chosen for Figure 4.4. The results for independently distributed marginal costs are found at $a = 0$.

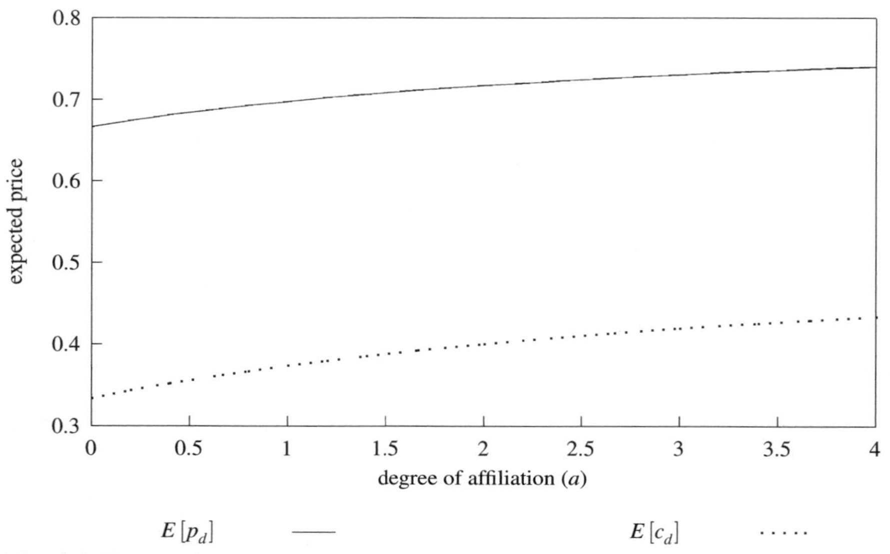

Fig. 4.4 Expected average price in a discriminatory auction (case A)

The graph of the expected average prices shows higher values for increasing a, i.e. for stronger affiliation. The relationship is the result of two effects. The first is the distribution itself. Because only the winner is paid, the probability of winning is important. The increasing effect of the affiliation is shown in Figure 3.3. Another reason is the bidding behavior. Given the demand range $(0,k]$, only the bid of the winner is relevant. Stronger affiliation leads to higher bids, see Figure 4.2. Both effects together point in the same direction, which strengthens the relationship. Therefore, consumers pay higher prices with affiliated marginal costs compared to independently distributed costs in the example.

Although the bidding behavior leads to lowest expected production costs, Figure 4.4 shows that the expected average price is higher than costs. This result is not surprising because only the winner earns something. He is paid his bid, which is strictly greater than his marginal costs. Hence, the average price is higher than the expected production costs because the probability distribution is the same. This relationship is stated in Corollary 4.3 (a).

4.3.2 Case B: Both Generators Are Necessary

Theorem 4.6. *Suppose $d \in (k, 2k)$. The unique Bayes-Nash equilibrium in symmetric and strictly increasing bidding strategies is given by $\left(b_d^*, b_d^*\right)$ with the following bid function*

$$b_d^*(c) = \left\{ \bar{p} - \bar{c} + \int_c^{\bar{c}} \exp\left(\int_\tau^{\bar{c}} \frac{[2k-d] f(t|t)}{k - [2k-d] F(t|t)} dt \right) d\tau \right\} \times \exp\left(- \int_c^{\bar{c}} \frac{[2k-d] f(t|t)}{k - [2k-d] F(t|t)} dt \right) + c \tag{4.82}$$

under the condition that

$$\frac{1}{b_d^*(c) - c} > \frac{f_i(c|c)}{f(c|c)} + \frac{[2k-d] F_i(c|c)}{k - [2k-d] F(c|c)} \tag{4.83}$$

holds true for all marginal costs, or, which is equivalent, under the condition that $\bar{p}_d$ exists and that

$$\bar{p} < \bar{p}_d. \tag{4.84}$$

$\bar{p}_d$ is the upper limit of the price cap and is obtained by

$$\bar{p}_d = \bar{c} + \min\left(\exp\left(\int_c^{\bar{c}} \frac{[2k-d] f(t|t)}{k - [2k-d] F(t|t)} dt \right) \times \left\{ \frac{1}{\frac{f_i(c|c)}{f(c|c)} + \frac{[2k-d] F_i(c|c)}{k - [2k-d] F(c|c)}} - \int_c^{\bar{c}} \exp\left(- \int_c^{\tau} \frac{[2k-d] f(t|t)}{k - [2k-d] F(t|t)} dt \right) d\tau \right\} \right) \tag{4.85}$$

for all marginal costs with

$$0 < \frac{f_i(c|c)}{f(c|c)} + \frac{[2k-d] F_i(c|c)}{k - [2k-d] F(c|c)}. \tag{4.86}$$

There exists no $\bar{p}_d$ if no marginal costs can be found that satisfy (4.86). The condition is always satisfied by independently distributed marginal costs. The optimal bid function for such distributed costs reduces to

$$b_d^{*i}(c) = \frac{[d-k][\bar{p}-c] + [2k-d]\int_c^{\bar{c}}[1-F(\tau)]\,d\tau}{k-[2k-d]F(c)} + c. \tag{4.87}$$

Proof.

The decision problem of a generator can be modeled equivalent to the problem in Theorem 4.2. Since the proof of Theorem 4.2 included the demand range $d \in (k, 2k)$ already, the results follow. See (4.51), (4.40), (4.53), (4.52), and (4.54). □

Some generators will deviate from the optimal bid function (4.82) if $\bar{p}_d$ exists and $\bar{p} \geq \bar{p}_d$. Hence, the auctioneer has to choose a price cap below $\bar{p}_d$ to ensure that all generators bid according to (4.82) if some marginal costs exist satisfying (4.86).

Knowing that both generators are essential to serve demand should reduce competitive pressure. A consequence should be that generators require higher payments. The question of whether lower marginal costs leads to a higher probability of winning is also answered by Corollary 4.4, among other properties of bidding behavior.

Corollary 4.4.

(a) *Given $d \in (k, 2k)$, bid shading exists, i.e. the optimal bid is greater than the marginal costs of the generator.*
(b) *Given $d \in (k, 2k)$, the optimal bid function increases strictly with marginal costs.*
(c) *Given $d \in (k, 2k)$, the optimal bid function increases strictly with demand for all marginal costs but $\bar{c}$, the highest one. The optimal bid for $\bar{c}$ is independent of demand.*
(d) *Optimal bids converge to bids for $d \in (0, k]$ when demand reaches k, the production capacity of a single generator.*
(e) *Optimal bids converge to price cap $\bar{p}$ when demand reaches $2k$, the production capacity of the whole market.*
(f) *Optimal bids are always higher for $d \in (k, 2k)$ compared to bids of demand range $d \in (0, k]$.*

Proof.

(a) Because the proof of Theorem 4.2 also incorporated the demand range $d \in (k, 2k)$, condition (4.33) also holds true in this case. It shows that bids have to be greater than marginal costs to maximize the expected profit and hence to represent the optimal bidding strategy.
(b) It was shown in the proof for Theorem 4.2 that the optimal bid exists for strictly increasing bids with respect to marginal costs, see (4.32). This proof incorporated the demand range $d \in (k, 2k)$. Hence, the result is applicable here.
(c) The first derivative of the bid function with respect to demand, using (4.82), is given by

$$\frac{\partial b_d^*(c)}{\partial d} = k[\bar{p}-\bar{c}]\mathrm{e}^{\int_c^{\bar{c}} \frac{(-)[2k-d]f(t|t)}{k-[2k-d]F(t|t)}dt} \int_c^{\bar{c}} \frac{f(t|t)}{\{k-[2k-d]F(t|t)\}^2}dt$$

$$+k\int_c^{\bar{c}} \mathrm{e}^{\int_c^{\tau} \frac{(-)[2k-d]f(t|t)}{k-[2k-d]F(t|t)}dt} \int_c^{\tau} \frac{f(t|t)}{\{k-[2k-d]F(t|t)\}^2}dt\,d\tau. \quad (4.88)$$

The derivative is zero for $\bar{c}$ because both terms on the right-hand side have integrals as factors, which become zero for the highest marginal costs. Both terms are positive for all other marginal costs because $\bar{p} > \bar{c}$, see Assumption 3.12, and $f(c|c) > 0\,\forall\, c \in [\underline{c}, \bar{c}]$, see Assumption 3.7. Hence, the optimal bid function strictly increases with demand for $c \in [\underline{c}, \bar{c})$ because the first derivative is positive. The bid function is independent of demand for $\bar{c}$ due to the slope of zero.

(d) Optimal bids of this case, see (4.82), converge for $d \longrightarrow k_+$ to

$$\lim_{d\to k_+} b_d^*(c) = \lim_{d\to k_+}\left\{\bar{p}-\bar{c}+\int_c^{\bar{c}} \mathrm{e}^{\int_\tau^{\bar{c}} \frac{[2k-d]f(t|t)}{k-[2k-d]F(t|t)}dt}d\tau\right\}\mathrm{e}^{\int_c^{\bar{c}} \frac{(-)[2k-d]f(t|t)}{k-[2k-d]F(t|t)}dt}+c$$

$$= \left[\bar{p}-\bar{c}+\int_c^{\bar{c}} \mathrm{e}^{\int_\tau^{\bar{c}} \frac{f(t|t)}{1-F(t|t)}dt}d\tau\right]\mathrm{e}^{\int_c^{\bar{c}} \frac{(-)f(t|t)}{1-F(t|t)}dt}+c.$$

It is the same expression as for the optimal bid function (4.18).

(e) Optimal bids for $d \in (k, 2k)$, see (4.82), converge for $d \longrightarrow 2k_-$ to

$$\lim_{d\to 2k_-} b_d^*(c) = \lim_{d\to 2k_-}\left\{\bar{p}-\bar{c}+\int_c^{\bar{c}} \mathrm{e}^{\int_\tau^{\bar{c}} \frac{[2k-d]f(t|t)}{k-[2k-d]F(t|t)}dt}d\tau\right\}\mathrm{e}^{\int_c^{\bar{c}} \frac{(-)[2k-d]f(t|t)}{k-[2k-d]F(t|t)}dt}+c$$

$$= \left[\bar{p}-\bar{c}+\int_c^{\bar{c}} 1\,d\tau\right]+c$$

$$= \bar{p}.$$

(f) Optimal bids of $d \in (k, 2k)$, see (4.82), converge at the lower boundary of their demand range to bids of case A, see (d). Hence, they are greater than those bids of $d \in (0, k]$, because (4.82) increase with demand due to (c) and these bids are defined for a demand higher than k. □

For economic reasons, lower marginal costs should result in lower bids. This will usually lead to the lowest production costs in the market. Corollary 4.4 (b) shows that the competitive advantage of having low costs pays off in form of a low bid. Unfortunately, the resulting bid does not show the true marginal costs. It is higher according to Corollary 4.4 (a). Hence, the paid price for electricity is always higher than the production costs because the offered bid price is paid in this auction type.

Bids have to be greater than marginal costs here because otherwise, according to the auction rules, generators would earn nothing or even lose money. Such high bids prevent the market from a failure due to a lack of supply.

The bidding behavior changes smoothly with increasing demand according to Corollary 4.4 (c). It shows that the bargaining power shifts with higher demand from consumers to generators. This is the result of the increasing fraction of assured production utilization expressed by $[d-k]$. This fraction becomes more valuable. Consequently, the risky amount of production capacity $[2k-d]$ decreases. A generator is now eager to get a higher price for the safe demand fraction. He can raise his bid slightly because his competitor faces the same situation. So far both competitors have the same intention and act in the same manner. It is not surprising, that bids converge to the highest possible bid $\bar{p}$ as Corollary 4.4 (e) states.

Furthermore, there exists no jump in the bid function on the border between case A and this case, see Corollary 4.4 (e). Hence, a small error in recognizing the true demand would not lead to a too great loss. Only the generator with the worst cost $\bar{c}$ will not make an error because his bid is independent of demand. It is the highest possible bid $\bar{p}$.

Example 4.5.

Suppose the probability distribution given by Definition 3.5. The optimal bid function for independently distributed marginal costs is given by

$$b_d^{*i}(c) = \frac{[d-k]\bar{p} + [2k-d]\frac{1-c^2}{2}}{k-[2k-d]c}. \tag{4.89}$$

For strong affiliation, the expression is obtained by

$$b_d^*(c) = \left\{ \bar{p} - 1 + \int_c^1 \exp\left(\int_\tau^1 \frac{2[2k-d]\left[1+at^2\right]}{k[2+at]-[2k-d][2+at^2]t} dt \right) d\tau \right\}$$
$$\times \exp\left(- \int_c^1 \frac{2[2k-d]\left[1+at^2\right]}{k[2+at]-[2k-d][2+at^2]t} dt \right) + c \tag{4.90}$$

if the following condition is satisfied for all $c > \frac{1-\sqrt{\frac{d}{k}-1}}{2-\frac{d}{k}}$

$$b_d^*(c) < \left[\frac{1}{a} + c^2\right] \frac{k[2+ac]-[2k-d][2+ac^2]c}{k[2c-1]-[2k-d]c^2} + c. \tag{4.91}$$

Proof.

The bid function for independently distributed marginal costs, see (4.87) and (3.25), is given by

$$b_d^{*i}(c) = \frac{[d-k][\bar{p}-c]+[2k-d]\int\limits_c^1[1-\tau]\,d\tau}{k-[2k-d]c}+c$$

$$= \frac{[d-k][\bar{p}-c]+[2k-d]\left[1-c-\frac{1-c^2}{2}\right]+kc-[2k-d]c^2}{k-[2k-d]c}$$

$$= \frac{[d-k]\bar{p}+[2k-d]\frac{1-c^2}{2}}{k-[2k-d]c}.$$

Given strong affiliation, the optimal bid, using (4.82), (3.26), and (3.27), is obtained by

$$b_d^*(c) = \left\{\bar{p}-1+\int\limits_c^1 \mathrm{e}^{\int\limits_\tau^1 \frac{[2k-d]\frac{1+at^2}{1+\frac{a}{2}t}}{k-[2k-d]\frac{1+\frac{a}{2}t^2}{1+\frac{a}{2}t}t}dt} d\tau\right\}\mathrm{e}^{\int\limits_c^1 \frac{(-)[2k-d]\frac{1+at^2}{1+\frac{a}{2}t}}{k-[2k-d]\frac{1+\frac{a}{2}t^2}{1+\frac{a}{2}t}t}dt}+c$$

$$= \left\{\bar{p}-1+\int\limits_c^1 \mathrm{e}^{\int\limits_\tau^1 \frac{2[2k-d]\left[1+at^2\right]}{k[2+at]-[2k-d]\left[2+at^2\right]t}dt} d\tau\right\}\mathrm{e}^{\int\limits_c^1 \frac{(-2)[2k-d]\left[1+at^2\right]}{k[2+at]-[2k-d]\left[2+at^2\right]t}dt}+c.$$

The bid function is only then optimal if condition (4.83) is satisfied. Using (3.26), (3.27), (3.31), and (3.32), it is obtained by

$$\frac{1}{b_d^*(c)-c} > \frac{f_i(c|c)}{f(c|c)}+\frac{[2k-d]F_i(c|c)}{k-[2k-d]F(c|c)}$$

$$> \frac{\frac{a\left[c-\frac{1}{2}\right]}{\left[1+\frac{a}{2}c\right]^2}}{\frac{1+ac^2}{1+\frac{a}{2}c}}-\frac{[2k-d]\frac{a[1-c]c}{2\left[1+\frac{a}{2}c\right]^2}}{k-[2k-d]\frac{1+\frac{a}{2}c^2}{1+\frac{a}{2}c}c}$$

$$> \frac{a[2c-1]}{[1+ac^2][2+ac]}-\frac{2[2k-d]a[1-c]c}{[2+ac]\{k[2+ac]-[2k-d][2+ac^2]c\}}$$

$$> \frac{a}{2+ac}\left\{\frac{2c-1}{1+ac^2}-\frac{2[2k-d][1-c]c}{k[2+ac]-[2k-d][2+ac^2]c}\right\}$$

$$> a\frac{[2c-1]\{k[2+ac]-[2k-d][2+ac^2]c\}-2[2k-d][1+ac^2][1-c]c}{[2+ac][1+ac^2]\{k[2+ac]-[2k-d][2+ac^2]c\}}$$

$$> a\frac{k[2c-1][2+ac]+[2k-d]c\{[2+ac^2][1-2c]-2[1+ac^2][1-c]\}}{[2+ac][1+ac^2]\{k[2+ac]-[2k-d][2+ac^2]c\}}$$

$$> a \frac{k[2c-1][2+ac]+[2k-d][1-2c]c-[2k-d][1+ac^2]c}{[2+ac][1+ac^2]\{k[2+ac]-[2k-d][2+ac^2]c\}}$$

$$> \frac{(-a)}{1+ac^2} \frac{k[1-2c]+[2k-d]c^2}{k[2+ac]-[2k-d][2+ac^2]c}. \tag{4.92}$$

The condition is always satisfied without calculating the optimal bid if the following holds true because bids are greater than marginal costs due to Corollary 4.4 (a), $a > 0$, and $c \geq 0$, see (3.16)

$$0 \geq \frac{(-a)}{1+ac^2} \frac{k[1-2c]+[2k-d]c^2}{k[2+ac]-[2k-d][2+ac^2]c}$$

$$\Longleftrightarrow$$

$$0 \leq \frac{k[1-2c]+[2k-d]c^2}{k[2+ac]-[2k-d][2+ac^2]c}$$

$$\Longleftrightarrow$$

$$0 < k[2+ac]-[2k-d][2+ac^2]c \tag{4.93}$$

and

$$0 \leq [2k-d]c^2 - k[2c-1]. \tag{4.94}$$

Inequation (4.93) is valid if the following condition holds true

$$k[2+ac] > [2k-d][2+ac^2]c$$

$$\Longleftrightarrow$$

$$(-2)\{k-[2k-d]c\} < ac\{k-[2k-d]c^2\}.$$

The factors in brackets are always positive because $[2k-d] < k$ and $c \in [0,1]$. Hence, the inequation is valid as well as (4.93). Taking into account $d \in (k, 2k)$, inequation (4.94) holds true if

$$0 \leq c^2 - \frac{2c}{2-\frac{d}{k}} + \frac{1}{2-\frac{d}{k}}$$

$$\leq \left[c - \frac{1}{2-\frac{d}{k}}\right]^2 - \frac{\frac{d}{k}-1}{\left[2-\frac{d}{k}\right]^2}$$

is valid. Because $d \in (k, 2k)$ and therefore $\frac{1}{2-\frac{d}{k}} > 1$, the condition is satisfied if the following holds true

$$\frac{\sqrt{\frac{d}{k}-1}}{2-\frac{d}{k}} \leq \frac{1}{2-\frac{d}{k}} - c \iff c \leq \frac{1-\sqrt{\frac{d}{k}-1}}{2-\frac{d}{k}}. \tag{4.95}$$

Bids for marginal costs, which do not satisfy (4.95), must be checked by (4.92). That condition can then be rewritten to

$$b_d^*(c) < \left[\frac{1}{a} + c^2\right] \frac{k[2+ac] - [2k-d][2+ac^2]c}{k[2c-1] - [2k-d]c^2} + c$$

because the right term of (4.92) is positive for $c > \frac{1-\sqrt{\frac{d}{k}-1}}{2-\frac{d}{k}}$. □

Bids for independently distributed marginal costs are shown in Figure 4.5. Results for strong affiliated marginal costs are presented in Figure 4.6 with $a = 4$. Bids increase as expected with higher marginal costs as well as with higher demand. Figure 4.7 shows the absolute bid difference between different degrees of affiliation. Negative values mean that bids for the independent distribution are higher than appropriate bids for strong affiliated costs.

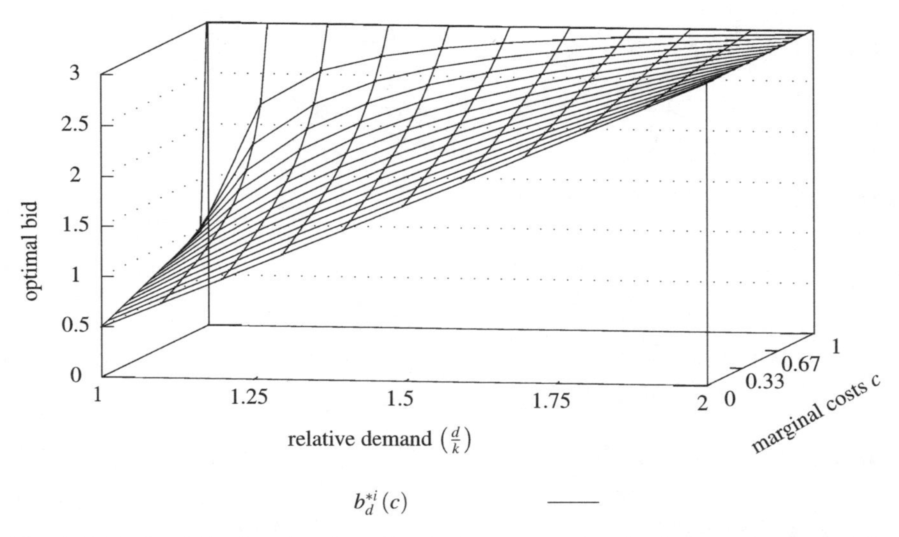

Fig. 4.5 Optimal bid function in a discriminatory auction for independently distributed costs (case B)

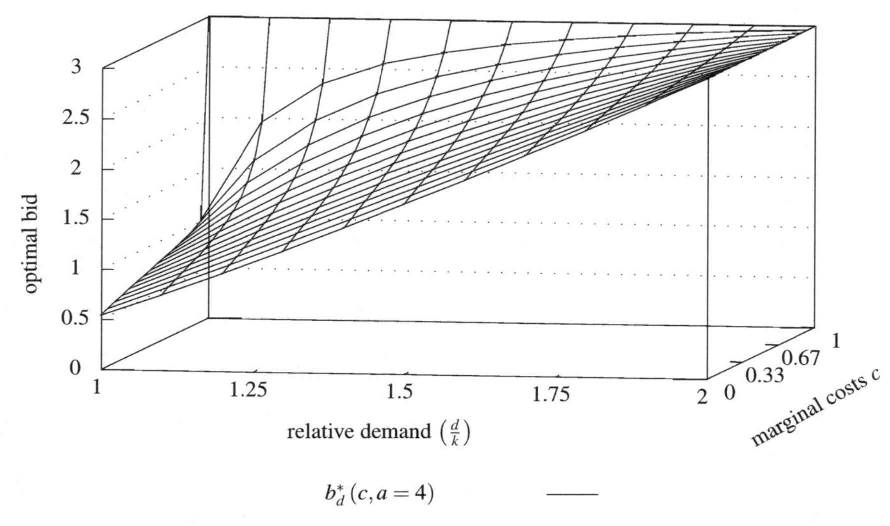

Fig. 4.6 Optimal bid function in a discriminatory auction for strong affiliation (case B)

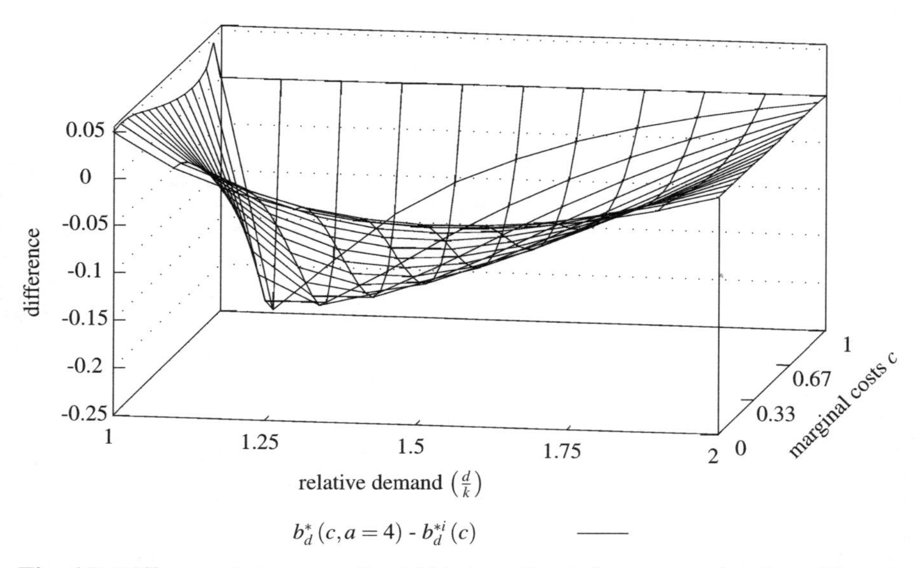

Fig. 4.7 Difference between optimal bids in a discriminatory auction (case B)

The question of how much higher a generator bids with strong affiliated costs relative to independently distributed costs is answered by Figure 4.8. For every demand $d \in (k, 2k)$, there exists at least one generator in the example who bids lower with strong affiliation. Some generators bid higher for $a = 4$ than with independently distributed costs. But they do so only up to a demand of $d \approx 1.175k$. The auctioneer sees therefore equal or lower bids for demands greater than about $1.175k$.

It seems that for this demand range a stronger affiliation is better for consumers because the bids are lower. But this conclusion is not correct because consumers are interested in the average price, which they expect to pay per electricity unit. This key figure is also influenced by the distribution of marginal costs itself. The single effect of the distribution may outweigh lower bids.

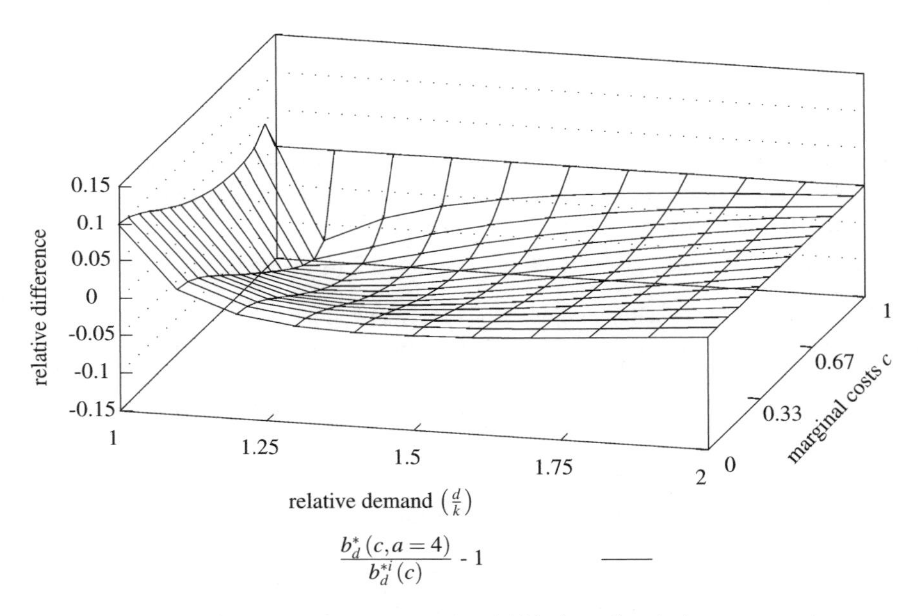

Fig. 4.8 Relative difference between optimal bids in a discriminatory auction (case B)

The results of Theorem 4.7 concerning expected production costs and efficiency rely on the general results of Theorem 4.1.

Theorem 4.7. *Suppose* $d \in (k, 2k)$.

(a) *The auction leads to production efficiency. Production costs are the lowest possible.*

(b) *The expected production costs per electricity unit are given by*

$$E\left[c_d\right] = 2\left[1 - \frac{k}{d}\right]\bar{c} - 2\frac{k}{d}\int\limits_{\underline{c}}^{\bar{c}} F(c)\,dc$$

$$+2\left[\frac{2k}{d} - 1\right]\int\limits_{\underline{c}}^{\bar{c}} F(c)\left\{c[f(c|c) + F_i(c|c)] + F(c|c)\right\}dc. \quad (4.96)$$

Given independently distributed marginal costs, the expression reduces to

$$E\left[c_d^i\right] = \bar{c} + \left[\frac{2k}{d} - 1\right]\int\limits_{\underline{c}}^{\bar{c}} F(c)^2\,dc - \frac{2k}{d}\int\limits_{\underline{c}}^{\bar{c}} F(c)\,dc. \quad (4.97)$$

Proof.

(a) Optimal bid function (4.82) is increasing in marginal costs due to Corollary 4.4 (b) and represents bids, which are accepted according to the auction rules. Hence, Theorem 4.1 applies, especially (a).

(b) Based on the same reason given in (a), (4.6) and (4.7) can be used. □

That generators are scheduled with lowest total production costs is not surprising. It is the result of raising bids with higher marginal costs. Hence, bids serve as an indicator of competitive advantage. This is good news because input factors are used in the most efficient way, presuming that input prices reflect market prices.

Theorem 4.8. *Suppose $d \in (k, 2k)$.*

(a) *The expected profit of a generator is obtained by*

$$\pi_d^*(c) = \{k - [2k - d]F(c|c)\}\exp\left(-\int\limits_{c}^{\bar{c}} \frac{[2k-d]f(t|t)}{k - [2k-d]F(t|t)}\,dt\right)$$

$$\times\left\{\bar{p} - \bar{c} + \int\limits_{c}^{\bar{c}} \exp\left(\int\limits_{\tau}^{\bar{c}} \frac{[2k-d]f(t|t)}{k - [2k-d]F(t|t)}\,dt\right)d\tau\right\}. \quad (4.98)$$

For independently distributed marginal costs, the expression reduces to

$$\pi_d^{*i}(c) = [d - k][\bar{p} - c] + [2k - d]\int\limits_{c}^{\bar{c}} [1 - F(\tau)]\,d\tau. \quad (4.99)$$

(b) *The expected contribution margin per capacity unit is given by*

$$\pi_d^{u^*}(c) = \left\{1 - \left[2 - \frac{d}{k}\right] F(c|c)\right\} \exp\left(-\int_c^{\bar{c}} \frac{[2k-d]f(t|t)}{k-[2k-d]F(t|t)} dt\right)$$
$$\times \left\{\bar{p} - \bar{c} + \int_c^{\bar{c}} \exp\left(\int_\tau^{\bar{c}} \frac{[2k-d]f(t|t)}{k-[2k-d]F(t|t)} dt\right) d\tau\right\}. \quad (4.100)$$

Given independently distributed marginal costs, the expression reduces to

$$\pi_d^{u^{*i}}(c) = \left[\frac{d}{k} - 1\right][\bar{p} - c] + \left[2 - \frac{d}{k}\right] \int_c^{\bar{c}} [1 - F(\tau)] d\tau. \quad (4.101)$$

Proof.

(a) Using optimal bid function (4.82), the expected profit of a generator (4.22), is determined by, taking into account (4.20), (4.24), (4.1), and (4.3)

$$\begin{aligned}
\pi_d^*(c) &= k[b(c) - c]y(c|c) \\
&= k\left[\bar{p} - \bar{c} + \int_c^{\bar{c}} e^{\int_\tau^{\bar{c}} \frac{(-)y_j(t|t)}{y(t|t)} dt} d\tau\right] e^{\int_c^{\bar{c}} \frac{y_j(t|t)}{y(t|t)} dt} y(c|c) \\
&= k[\alpha - \beta_f F(c|c)]\left[\bar{p} - \bar{c} + \int_c^{\bar{c}} e^{\int_\tau^{\bar{c}} \frac{\beta_f f(t|t)}{\alpha - \beta_f F(t|t)} dt} d\tau\right] e^{\int_c^{\bar{c}} \frac{(-)\beta_f f(t|t)}{\alpha - \beta_f F(t|t)} dt} \\
&= \{k - [2k-d]F(c|c)\}\left\{\bar{p} - \bar{c} + \int_c^{\bar{c}} e^{\int_\tau^{\bar{c}} \frac{[2k-d]f(t|t)}{k-[2k-d]F(t|t)} dt} d\tau\right\} \\
&\quad \times e^{\int_c^{\bar{c}} \frac{(-)[2k-d]f(t|t)}{k-[2k-d]F(t|t)} dt}.
\end{aligned}$$

For independently distributed marginal costs, the expected profit, using (3.11), (3.12), and (3.3), is obtained by

$$\begin{aligned}
\pi_d^{*i}(c) &= \{k - [2k-d]F(c)\}\left\{\bar{p} - \bar{c} + \int_c^{\bar{c}} e^{\int_\tau^{\bar{c}} \frac{[2k-d]f(t)}{k-[2k-d]F(t)} dt} d\tau\right\} e^{\int_c^{\bar{c}} \frac{(-)[2k-d]f(t)}{k-[2k-d]F(t)} dt} \\
&= [d-k][\bar{p} - \bar{c}] + \int_c^{\bar{c}} \{k - [2k-d]F(\tau)\} d\tau \\
&= [d-k]\bar{p} + [2k-d]\bar{c} - kc - [2k-d]\int_c^{\bar{c}} F(\tau) d\tau \\
&= [d-k][\bar{p} - c] + [2k-d]\int_c^{\bar{c}} [1 - F(\tau)] d\tau.
\end{aligned}$$

(b) The expected contribution margin per electricity unit, using (4.60), is given by

$$\pi_d^{u^*}(c) = \frac{\pi_d^*(c)}{k}$$
$$= \left\{1 - \left[2 - \frac{d}{k}\right] F(c|c)\right\} \left\{\bar{p} - \bar{c} + \int_c^{\bar{c}} e^{\int_\tau^{\bar{c}} \frac{[2k-d]f(t|t)}{k-[2k-d]F(t|t)} dt} d\tau\right\}$$
$$\times e^{\int_c^{\bar{c}} \frac{(-)[2k-d]f(t|t)}{k-[2k-d]F(t|t)} dt}.$$

Using (4.99), the expression for independently distributed marginal costs is obtained by

$$\pi_d^{u^*i}(c) = \left[\frac{d}{k} - 1\right][\bar{p} - c] + \left[2 - \frac{d}{k}\right] \int_c^{\bar{c}} [1 - F(\tau)]\, d\tau.$$

□

Corollary 4.5. *Suppose $d \in (k, 2k)$.*

(a) *The expected profit and expected contribution margin per capacity unit are positive.*

(b) *The expected profit and expected contribution margin per capacity unit increase strictly with demand.*

(c) *The expected profit and expected contribution margin per capacity unit decrease strictly with marginal costs if the following condition holds true for all marginal costs:*

$$\frac{1}{b_d^*(c) - c} > (-)\frac{[2k-d]F_i(c|c)}{k - [2k-d]F(c|c)}. \tag{4.102}$$

This condition is always satisfied for independent distribution.

Proof.

(a) The expressions of the expected profit, see (4.98) is positive because the first factor is greater than zero due to $d > k$. The second factor is the difference between optimal bid minus the marginal costs or the bid shading. It is positive due to Corollary 4.4 (a). Hence, the expected profit is positive. The same argument applies to the expected contribution margin because the additional factor $\frac{1}{k}$ is positive.

(b) The first derivative of the expected profit with respect to demand, using (4.98), is given by

$$\frac{\partial \pi_d^*(c)}{\partial d} = F(c|c)\left\{\bar{p}-\bar{c}+\int_c^{\bar{c}} \mathrm{e}^{\int_\tau^{\bar{c}} \frac{[2k-d]f(t|t)}{k-[2k-d]F(t|t)}dt} d\tau\right\} \mathrm{e}^{\int_c^{\bar{c}} \frac{(-)[2k-d]f(t|t)}{k-[2k-d]F(t|t)}dt}$$

$$+k\{k-[2k-d]F(c|c)\}\left([\bar{p}-\bar{c}]\mathrm{e}^{\int_c^{\bar{c}} \frac{(-)[2k-d]f(t|t)}{k-[2k-d]F(t|t)}dt}\right.$$

$$\times\int_c^{\bar{c}} \frac{f(t|t)}{\{k-[2k-d]F(t|t)\}^2}dt+\int_c^{\bar{c}} \mathrm{e}^{\int_c^{\tau} \frac{(-)[2k-d]f(t|t)}{k-[2k-d]F(t|t)}dt}$$

$$\left.\times\int_c^{\tau} \frac{f(t|t)}{\{k-[2k-d]F(t|t)\}^2}dt\,d\tau\right). \tag{4.103}$$

A simpler expression for the proof, using (4.82) and (4.88), is obtained by

$$\frac{\partial \pi_d^*(c)}{\partial d} = F(c|c)[b_d^*(c)-c]+\{k-[2k-d]F(c|c)\}\frac{\partial b_d^*(c)}{\partial d}.$$

Now it is easy to see that the first summand is positive for $c \in (c,\bar{c}]$ because $F(c|c) > 0$ and $b_d^*(c) > c$, see Corollary 4.4 (a). The second summand is non-negative for the same cost range due to $k > [2k-d]F(c|c)$ and because the derivative of the bid function is non-negative according to Corollary 4.4 (c). Given the lowest marginal cost $\underline{c}$, the first summand is obviously zero because $F(\underline{c}|\underline{c}) = 0$. The second summand is then positive due to Corollary 4.4 (c). Hence, the sum is always greater than zero. This means the expected profit increases strictly with demand.

The slope for the expected contribution margin, using (4.100), is obtained by

$$\frac{\partial \pi_d^{u^*}(c)}{\partial d} = \frac{1}{k}\frac{\partial \pi_d^*(c)}{\partial d}. \tag{4.104}$$

The derivative is also positive because of the same argument as for the expected profit.

(c) The expected profit strictly decreases with marginal costs if the first derivative with respect to marginal costs is negative. The first derivative, using (4.98) and (4.82), is given by

$$\frac{\partial \pi_d^*(c)}{\partial c} = \frac{\partial}{\partial c}\{k-[2k-d]F(c|c)\}[b_d^*(c)-c]$$

$$= (-)[2k-d][f(c|c)+F_i(c|c)][b_d^*(c)-c]$$

$$+\{k-[2k-d]F(c|c)\}\left[b_d^*(c)'-1\right].$$

Using (4.31), (4.24), (4.20), and (4.1)–(4.3), the term is rewritten to

$$\begin{aligned}\frac{\partial \pi_d^*(c)}{\partial c} &= (-)[2k-d][f(c|c)+F_i(c|c)][b_d^*(c)-c]-k+[2k-d]F(c|c)\\&\quad+[2k-d]f(c|c)[b_d^*(c)-c]\\&=[2k-d]F(c|c)-k-[2k-d]F_i(c|c)[b_d^*(c)-c].\end{aligned}$$

The first derivative of the expected profit is negative if the following holds true, taking into account Corollary 4.4 (a) and $F_i(c|c) \leq 0$ due to (3.9)

$$\frac{\partial \pi_d^*(c)}{\partial c} < 0$$

$$\Longleftrightarrow$$

$$k-[2k-d]F(c|c) > (-)[2k-d]F_i(c|c)[b_d^*(c)-c]$$

$$\Longleftrightarrow$$

$$\frac{1}{b_d^*(c)-c} > \frac{(-)[2k-d]F_i(c|c)}{k-[2k-d]F(c|c)}. \tag{4.105}$$

The condition for independently distributed marginal costs is given by, taking into account that $F_i(c|c)=0$, see (3.14), and $F(c|c) \leq 1$, see (3.3)

$$\frac{1}{b_d^*(c)-c} > 0.$$

This is valid because of Corollary 4.4 (a). Hence, the expected profit decreases strictly with marginal costs for the independent distribution.

Similar arguments are found for the expected contribution margin because the factor $\frac{1}{k}$ is independent of c. □

The expected profit is positive, see Corollary 4.5 (a), because the loser sells something and earns money. Hence, his profit is not zero. That the expected profit and the contribution margin strictly increase with demand is the result of a higher assured sold output in the case of losing the auction. The effect is strengthened by increasing optimal bids for $c \in [\underline{c}, \bar{c})$, see Corollary 4.4 (c).

Example 4.6. *Suppose the probability distribution given by Definition 3.5.*

(a) *The expected profit for independently distributed marginal costs is obtained by*

$$\pi_d^{*i}(c) = [d-k][\bar{p}-c] + [2k-d]\frac{[1-c]^2}{2}. \tag{4.106}$$

For strong affiliated costs with $a > 0$, the expression is given by

$$\pi_d^*(c) = \left\{k - [2k-d]\frac{2+ac^2}{2+ac}c\right\}$$
$$\times \left\{\bar{p} - 1 + \int_c^1 \exp\left(\int_\tau^1 \frac{2[2k-d]\left[1+at^2\right]}{k[2+at]-[2k-d]\left[2+at^2\right]t}\,dt\right)d\tau\right\}$$
$$\times \exp\left(-\int_c^1 \frac{2[2k-d]\left[1+at^2\right]}{k[2+at]-[2k-d]\left[2+at^2\right]t}\,dt\right). \tag{4.107}$$

(b) *The expected contribution margin per capacity unit for independently distributed costs is given by*

$$\pi_d^{u*i}(c) = \left[\frac{d}{k}-1\right][\bar{p}-c] + \left[2-\frac{d}{k}\right]\frac{[1-c]^2}{2}. \tag{4.108}$$

The expression for strong affiliation with $a > 0$ is obtained by

$$\pi_d^{u*}(c) = \left\{1 - \left[2-\frac{d}{k}\right]\frac{2+ac^2}{2+ac}c\right\}$$
$$\times \left\{\bar{p} - 1 + \int_c^1 \exp\left(\int_\tau^1 \frac{2[2k-d]\left[1+at^2\right]}{k[2+at]-[2k-d]\left[2+at^2\right]t}\,dt\right)d\tau\right\}$$
$$\times \exp\left(-\int_c^1 \frac{2[2k-d]\left[1+at^2\right]}{k[2+at]-[2k-d]\left[2+at^2\right]t}\,dt\right). \tag{4.109}$$

Proof.

(a) The expected profit of a generator (4.98), using (3.26) and (3.27), is given by

$$\pi_d^*(c) = \left\{k - [2k-d]\frac{1+\frac{a}{2}c^2}{1+\frac{a}{2}c}c\right\}\left\{\bar{p} - 1 + \int_c^1 e^{\int_\tau^1 \frac{[2k-d]\frac{1+at^2}{1+\frac{a}{2}t}}{k-[2k-d]\frac{1+\frac{a}{2}t^2}{1+\frac{a}{2}t}t}dt} d\tau\right\}$$

$$\times e^{\int_c^1 \frac{(-)[2k-d]\frac{1+at^2}{1+\frac{a}{2}t}}{k-[2k-d]\frac{1+\frac{a}{2}t^2}{1+\frac{a}{2}t}t}dt}$$

$$= \left\{k - [2k-d]\frac{2+ac^2}{2+ac}c\right\}\left\{\bar{p} - 1 + \int_c^1 e^{\int_\tau^1 \frac{2[2k-d]\left[1+at^2\right]}{k[2+at]-[2k-d]\left[2+at^2\right]t}dt} d\tau\right\}$$

$$\times e^{\int_c^1 \frac{(-2)[2k-d]\left[1+at^2\right]}{k[2+at]-[2k-d]\left[2+at^2\right]t}dt}.$$

The expression for independently distributed marginal costs, using (4.99) and (3.25), is obtained by

$$\pi_d^{*i}(c) = [d-k][\bar{p}-c] + [2k-d]\int_c^1 [1-\tau]\,d\tau$$

$$= [d-k][\bar{p}-c] + [2k-d]\left[1 - c - \left.\frac{\tau^2}{2}\right|_c^1\right]$$

$$= [d-k][\bar{p}-c] + [2k-d]\frac{[1-c]^2}{2}.$$

(b) The expected contribution margin per capacity unit (4.100), using (4.107), is given by

$$\pi_d^{u^*}(c) = \frac{\pi_d^*(c)}{k}$$

$$= \left\{1 - \left[2 - \frac{d}{k}\right]\frac{2+ac^2}{2+ac}c\right\} e^{\int_c^1 \frac{(-2)[2k-d]\left[1+at^2\right]}{k[2+at]-[2k-d]\left[2+at^2\right]t}dt}$$

$$\times \left\{\bar{p} - 1 + \int_c^1 e^{\int_\tau^1 \frac{2[2k-d]\left[1+at^2\right]}{k[2+at]-[2k-d]\left[2+at^2\right]t}dt} d\tau\right\}.$$

For independent distribution, the expression, using (4.106), is obtained by

$$\pi_d^{u*i}(c) = \left[\frac{d}{k} - 1\right][\bar{p} - c] + \left[2 - \frac{d}{k}\right]\frac{[1-c]^2}{2}. \qquad \square$$

Two effects are responsible for increased expected profits and contribution margins compared to case A. Firstly, less aggressive bidding with higher bids leads to higher prices for each electricity unit sold. On the other hand, utilization of production capacity rises. The reason is that now the loser also sells something. His fraction increases with demand. The whole effect is shown in Figure 4.9 for independently distributed marginal costs and in Figure 4.10 for strong affiliation with $a = 4$. The expected contribution margin declines in the example with higher marginal costs for both distributions.

The difference between the profit for one electricity unit for the strong affiliation case compared with the independent distribution is shown in Figure 4.11. How much more a generator can expect to earn relative to the case of independently distributed marginal costs is presented by Figure 4.12. Because the expected profit disappears for marginal costs close to $\bar{c}$, the relative difference is very high. The excess profit for strong affiliation in the example is for a demand greater than $1.05k$ yet below 15% of its value for independently distributed marginal costs. Some generators with strong affiliated cost face a lower expected profit. The loss does not exceed 2.5% of the earnings for independent distribution.

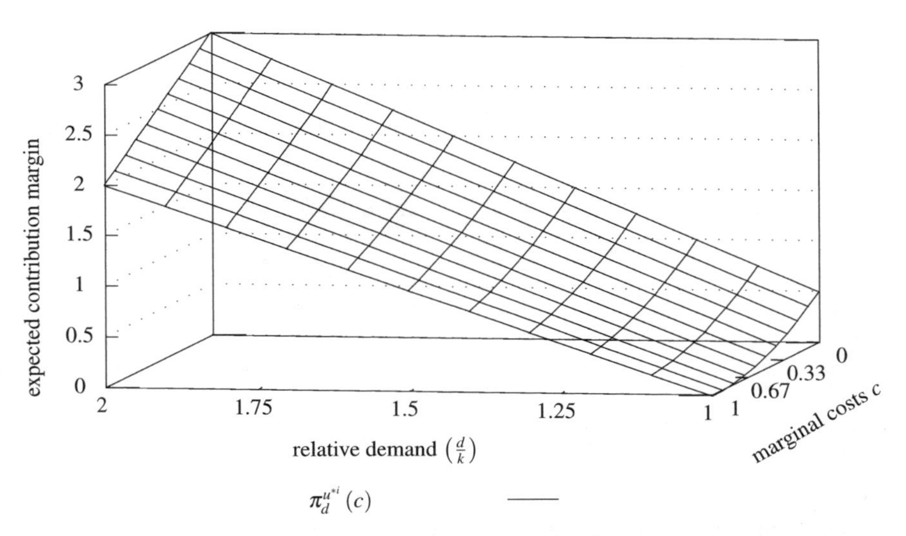

Fig. 4.9 Expected contribution margin in a discriminatory auction for independently distributed costs (case B)

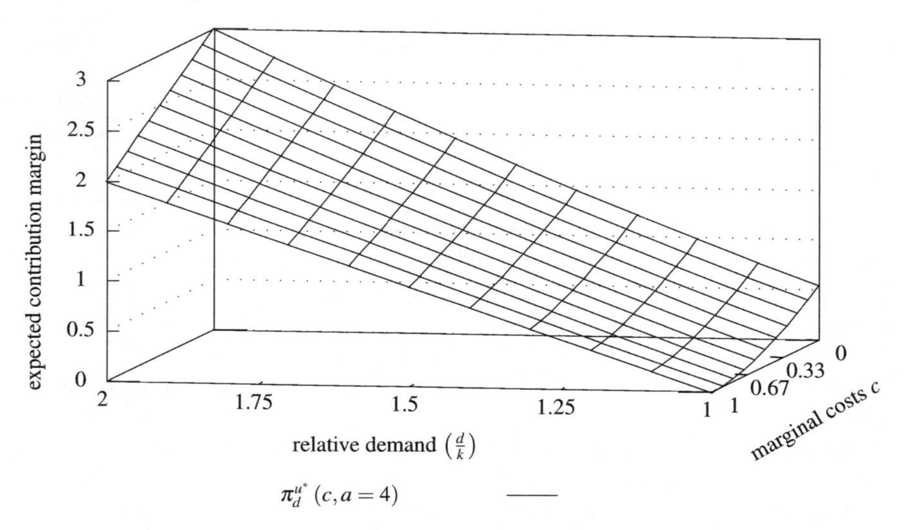

Fig. 4.10 Expected contribution margin in a discriminatory auction for strong affiliation (case B)

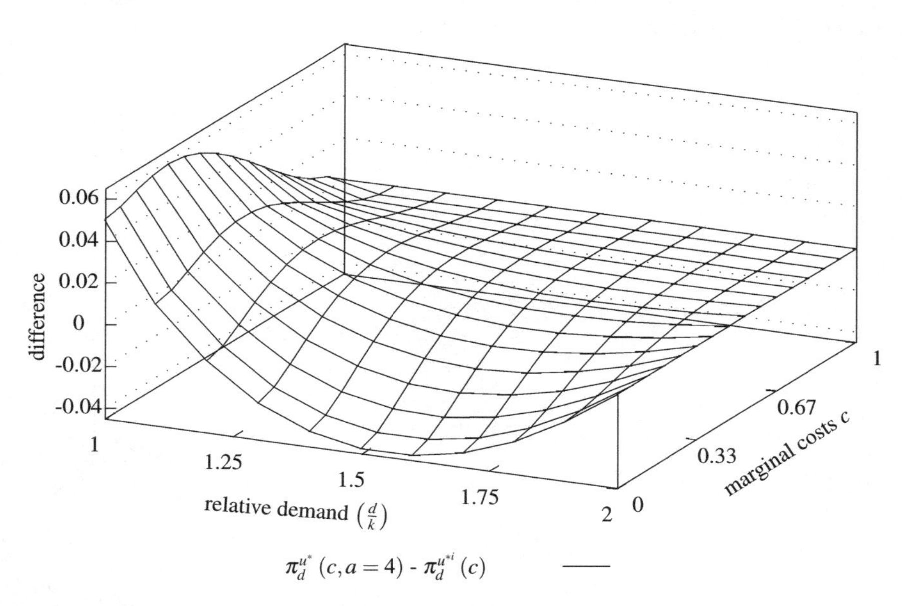

Fig. 4.11 Difference in contribution margins in a discriminatory auction (case B)

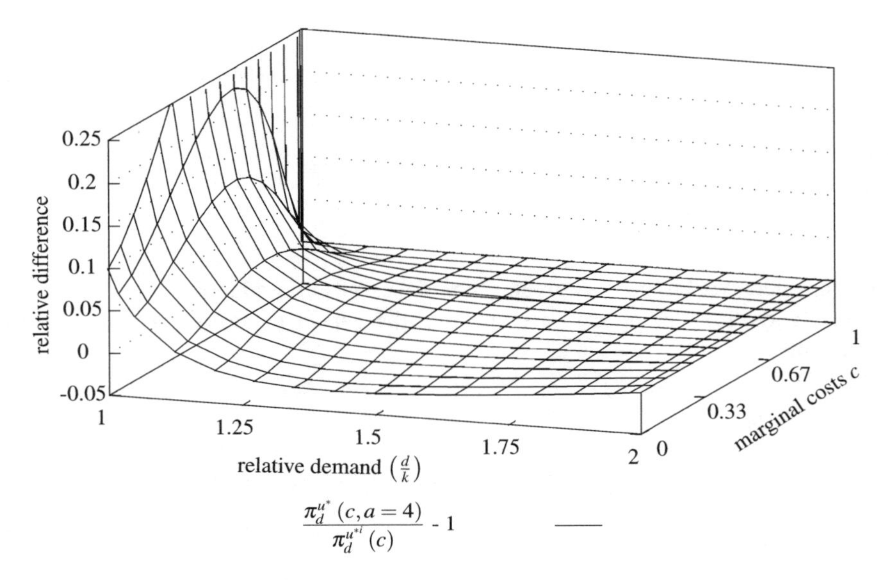

Fig. 4.12 Relative difference in contribution margins in a discriminatory auction (case B)

Theorem 4.9. *Suppose $d \in (k, 2k)$. The expected average price is obtained by*

$$E[p_d] = 2\left[1-\frac{k}{d}\right]\bar{p}+2\left[\frac{2k}{d}-1\right]\left(\int_{\underline{c}}^{\bar{c}} cF(c)\left[f(c|c)+F_i(c|c)\right]dc\right.$$
$$+\int_{\underline{c}}^{\bar{c}} F(c)F_i(c|c)\exp\left(-\int_{c}^{\bar{c}}\frac{[2k-d]f(t|t)}{k-[2k-d]F(t|t)}dt\right)\left\{\bar{p}-\bar{c}\right.$$
$$\left.\left.+\int_{c}^{\bar{c}}\exp\left(\int_{\tau}^{\bar{c}}\frac{[2k-d]f(t|t)}{k-[2k-d]F(t|t)}dt\right)d\tau\right\}dc\right). \tag{4.110}$$

For independently distributed marginal costs, the expression reduces to

$$E\left[p_d^i\right] = 2\left[1-\frac{k}{d}\right]\bar{p}+\left[\frac{2k}{d}-1\right]\left[\bar{c}-\int_{\underline{c}}^{\bar{c}} F(c)^2\,dc\right]. \tag{4.111}$$

Proof.

The expression has already been derived with the proof of Theorem 4.5, see (4.76) and (4.77). □

Corollary 4.6. *Suppose $d \in (k, 2k)$. The expected average price is greater than the expected production costs.*

Proof.

The expected average price is greater than the expected production costs if the following holds true, taking into account that the expected production costs (4.96) equals (4.6)

$$0 < E[p_d] - E[c].$$

Based on (4.75), the expected average price (4.73), using (4.1)– (4.3) and (4.20), is rewritten to

$$E[p_d] = \frac{2}{d} \int_{\underline{c}}^{\bar{c}} b_d^*(c) f(c) \{k - [2k-d] F(c|c)\} \, dc.$$

Based on (4.9), the lowest expected cost (4.6), using (4.1), (4.2), is given by

$$E[c] = 2\frac{2k-d}{d} \int_{\underline{c}}^{\bar{c}} cf(c) [1 - F(c|c)] \, dc + 2\frac{d-k}{d} \int_{\underline{c}}^{\bar{c}} cf(c) \, dc. \tag{4.112}$$

The average price is therefore greater than the production costs if

$$0 < \frac{2}{d} \left(\int_{\underline{c}}^{\bar{c}} b_d^*(c) f(c) \{k - [2k-d] F(c|c)\} \, dc - [d-k] \int_{\underline{c}}^{\bar{c}} cf(c) \, dc \right.$$
$$\left. - [2k-d] \int_{\underline{c}}^{\bar{c}} cf(c) [1 - F(c|c)] \, dc \right)$$

$$\Longleftrightarrow$$

$$0 < \int_{\underline{c}}^{\bar{c}} [b_d^*(c) - c] f(c) \{k - [2k-d] F(c|c)\} \, dc$$

is valid. It is sufficient that following holds true, because $d \in (k, 2k)$ and taking into account (3.5) as well as Assumption 3.7, declaring that the probability density function is positive for all marginal costs

$$b_d^*(c) > c.$$

This relation was proved by Corollary 4.4 (a). Therefore, the expected average price is greater than the expected production costs. □

The expected average price exceeds the expected production costs, see Corollary 4.6. It ensures that generators can earn positive profits and therefore stay in the market. Otherwise, the market would break down because nobody would produce electricity.

Example 4.7.

Suppose the probability distribution given by Definition 3.5. The expected average price for independently distributed marginal costs is obtained by

$$E\left[p_d^i\right] = 2\left[1-\frac{k}{d}\right]\bar{p}+\frac{2}{3}\left[\frac{2k}{d}-1\right]. \tag{4.113}$$

Given strong affiliation, the expression is

$$\begin{aligned} E\left[p_d\right] = 2\left[1-\frac{k}{d}\right]\bar{p}+\left[\frac{2k}{d}-1\right]\Bigg(\frac{4}{5}-\frac{2}{4+a}\Bigg[\frac{1}{10}+\frac{2}{3a}-\frac{4}{a^2}-\frac{16}{a^3} \\ +16\frac{2+a}{a^4}\ln\left(1+\frac{a}{2}\right)\Bigg]-\frac{4a}{4+a}\int_0^1\frac{4+ac}{[2+ac]^2}[1-c]c^2 \\ \times\left\{\bar{p}-1+\int_c^1\exp\left(\int_\tau^1\frac{2[2k-d]\left[1+at^2\right]}{k[2+at]-[2k-d]\left[2+at^2\right]t}dt\right)d\tau\right\} \\ \times\exp\left(-\int_c^1\frac{2[2k-d]\left[1+at^2\right]}{k[2+at]-[2k-d]\left[2+at^2\right]t}dt\right)dc\Bigg). \end{aligned} \tag{4.114}$$

Proof.

The expected average price for independently distributed marginal costs, using (4.111) and (3.25), is given by

$$
\begin{aligned}
E\left[p_d^i\right] &= 2\left[1-\frac{k}{d}\right]\bar{p}+\left[\frac{2k}{d}-1\right]\left[1-\int_0^1 c^2\,dc\right] \\
&= 2\left[1-\frac{k}{d}\right]\bar{p}+\frac{2}{3}\left[\frac{2k}{d}-1\right].
\end{aligned}
$$

According to (4.110), (3.25), (3.26), (3.27), (3.32), and (4.17), the expected average price for strong affiliation with $a > 0$ is obtained by

$$
\begin{aligned}
E\left[p_d\right] &= 2\left[1-\frac{k}{d}\right]\bar{p}+\left[\frac{2k}{d}-1\right]\Bigg(\frac{4}{5}-\frac{2}{4+a}\left[\frac{1}{10}+\frac{2}{3a}-\frac{4}{a^2}-\frac{16}{a^3}\right. \\
&\quad \left.+16\,\frac{2+a}{a^4}\ln\left(1+\frac{a}{2}\right)\right]-a\int_0^1 \frac{1+\frac{a}{4}c}{1+\frac{a}{4}}\,\frac{1-c}{\left[1+\frac{a}{2}c\right]^2}\,c^2 \\
&\quad \times\left\{\bar{p}-1+\int_c^1 \mathrm{e}^{\int_\tau^1 \frac{[2k-d]\frac{1+at^2}{1+\frac{a}{2}t}}{k-[2k-d]\frac{1+\frac{a}{2}t^2}{1+\frac{a}{2}t}t}\,dt}\,d\tau\right\}\mathrm{e}^{\int_c^1 \frac{(-)[2k-d]\frac{1+at^2}{1+\frac{a}{2}t}}{k-[2k-d]\frac{1+\frac{a}{2}t^2}{1+\frac{a}{2}t}t}\,dt}\,dc\Bigg) \\
&= 2\left[1-\frac{k}{d}\right]\bar{p}+\left[\frac{2k}{d}-1\right]\Bigg(\frac{4}{5}-\frac{2}{4+a}\left[\frac{1}{10}+\frac{2}{3a}-\frac{4}{a^2}-\frac{16}{a^3}\right. \\
&\quad \left.+16\,\frac{2+a}{a^4}\ln\left(1+\frac{a}{2}\right)\right]-\frac{4a}{4+a}\int_0^1 \frac{4+ac}{[2+ac]^2}[1-c]c^2 \\
&\quad \times\left\{\bar{p}-1+\int_c^1 \mathrm{e}^{\int_\tau^1 \frac{2[2k-d]\left[1+at^2\right]}{k[2+at]-[2k-d]\left[2+at^2\right]t}\,dt}\,d\tau\right\} \\
&\quad \times\mathrm{e}^{\int_c^1 \frac{(-2)[2k-d]\left[1+at^2\right]}{k[2+at]-[2k-d]\left[2+at^2\right]t}\,dt}\,dc\Bigg).
\end{aligned}
$$

□

Figure 4.13 shows the expected average prices for various degrees of affiliation. The average price increases with higher demand in the example. Figure 4.14 shows the expected average price and the appropriate expected production costs for a selection of demands. The average price is much larger than the expected costs.

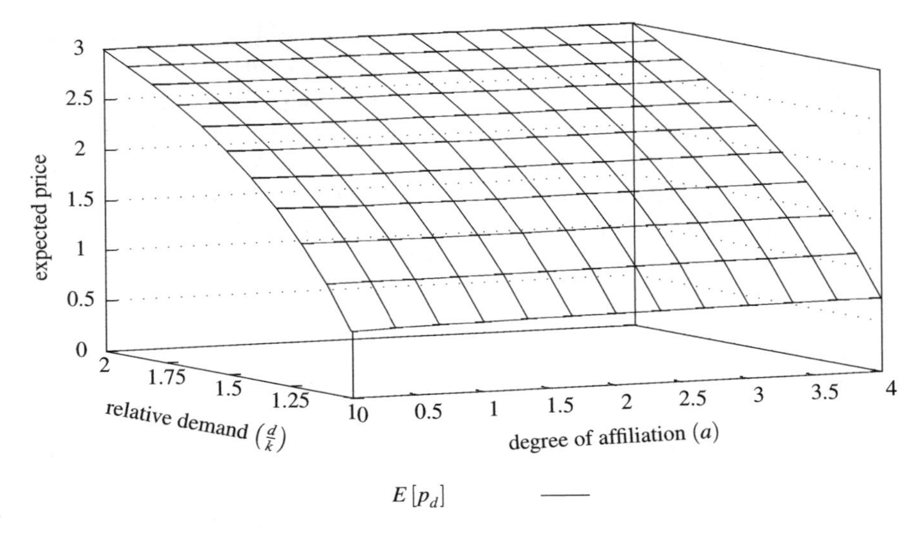

Fig. 4.13 Expected average price in a discriminatory auction (case B)

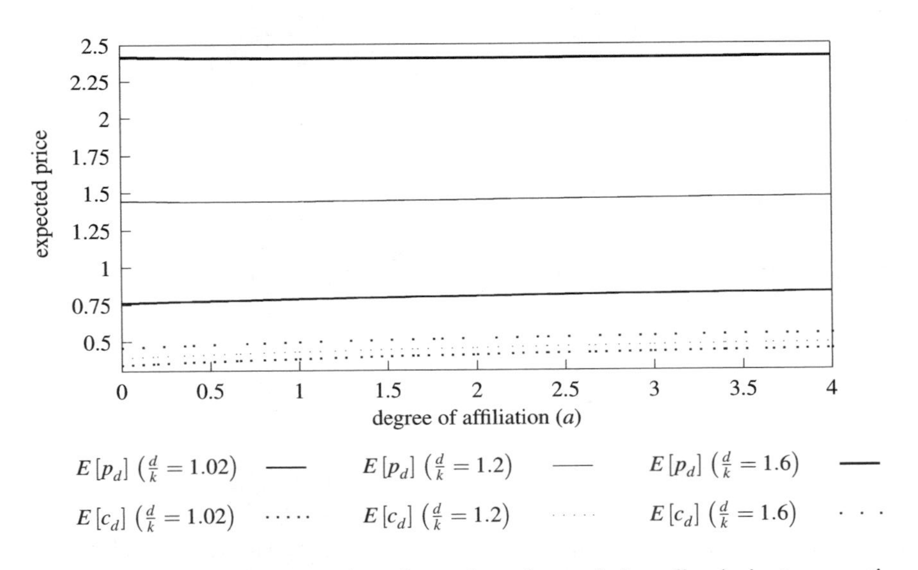

Fig. 4.14 Expected average prices for various demands in a discriminatory auction (case B)

The excess average price for strong affiliation is shown in Figure 4.15. The expected average price for strong affiliation does not exceed the value for the independent distribution from a demand of $\approx 1.45k$ on. Consumers save less than 0.6% of the expected average price in this demand range. On the other hand, consumers expect higher average prices for strong affiliation up to a demand of $\approx 1.1k$. The picture is mixed for the remaining demand range from $\approx 1.1k$ to $\approx 1.45k$. Depending on the degree of affiliation, the average price can be lower or higher compared with the appropriate value for independent distribution.

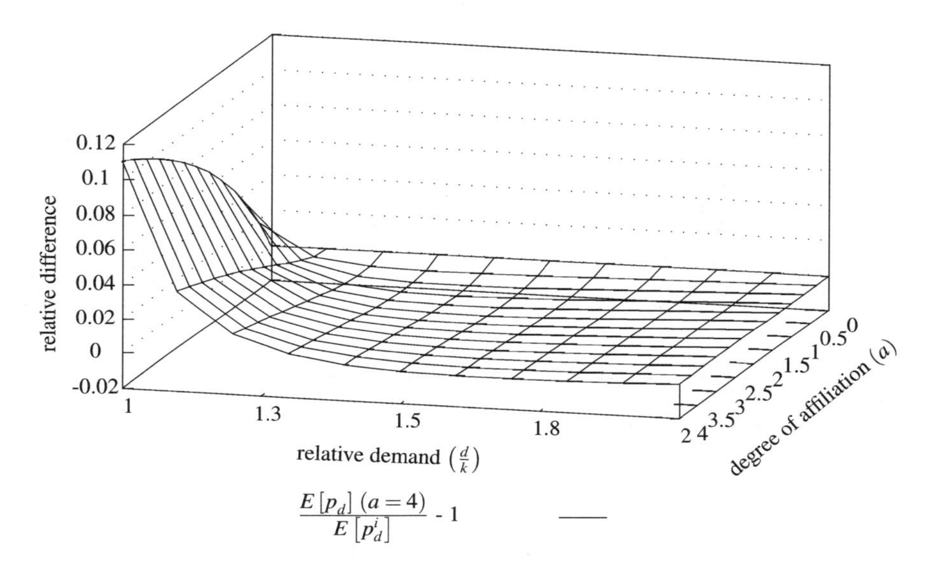

Fig. 4.15 Relative difference between expected average prices in a discriminatory auction (case B)

4.3.3 Case C: Full Market Capacity Utilization

Generators know in case C, with $d \geq 2k$, that all of them are necessary. The system operator must ration demand if it exceeds the market production capacity. Nevertheless, succeeding in the auction still depends on bids, but the focus is set on the price cap rule. It avoids the bankruptcy of the consumers with reservation price $\bar{p}$. The Bayes–Nash equilibrium and remaining key figures are given in Theorem 4.10.

Theorem 4.10. *Suppose* $d \geq 2k$.

(a) *The unique Bayes–Nash equilibrium is given by* $\left(b_d^*, b_d^*\right)$ *with the following bid function*

$$b_d^*(c) = \bar{p}. \tag{4.115}$$

(b) *The auction leads to production efficiency with the lowest possible production costs.*

(c) *The expected costs per electricity unit are given by*

$$E[c_d] = \bar{c} - \int_{\underline{c}}^{\bar{c}} F(c)\,dc. \tag{4.116}$$

(d) *The profit is determined by*

$$\pi_d^*(c) = k[\bar{p} - c]. \tag{4.117}$$

(e) *The contribution margin per capacity unit is given by*

$$\pi_d^{u^*}(c) = \bar{p} - c. \tag{4.118}$$

(f) *The expected average price is given by*

$$E[p_d] = \bar{p}. \tag{4.119}$$

Proof.

(a) Every generator can sell his whole production capacity k because demand is high enough, due to $d \geq 2k$, so that no generator has unused production capacity. Presuming the bid is accepted by the auctioneer, the output is independent of the bid. Hence, only the contribution margin determines the profit because the price paid equals the bid. Maximizing the profit equals maximizing the contribution margin. It is highest with the highest accepted bid. This is $\bar{p}$ according to the price cap rule, see Assumption 3.12. Notice, the bid is independent of marginal costs.

(b) Because the requirements of Theorem 4.1 are satisfied due to optimal bidding $\bar{p}$ and demand $d \geq 2k$, the result of Theorem 4.1 (a) applies.

(c) Theorem 4.1 (b) applies because of the same argument given in (b). Equation (4.8) is used because demand is greater than or equal to the aggregated production capacity.

$$E[c_d] = E[c] = \bar{c} - \int_{\underline{c}}^{\bar{c}} F(\tau)\,d\tau.$$

(d) This follows directly from the model because the whole production capacity k of a generator is sold and the price paid is always $\bar{p}$. Hence, the profit is simply

$$\pi_d^*(c) = k[\bar{p} - c].$$

(e) The contribution margin for each electricity unit, see (4.117), is

$$\pi_d^{u^*}(c) = \frac{\pi_d^*(c)}{k} = \bar{p} - c.$$

(f) This is obvious because all generators bid $\bar{p}$. □

The results are not surprising. If both generators use their full production capacity then they require the highest possible price. Consumers are now in their worst bargaining position because the generators act like one entity. This eliminates possible competitive advantages by playing one generator against the other.

Example 4.8. *Suppose the probability distribution given by Definition 3.5.*

(a) *The optimal bid function is given by*

$$b_d^*(c) = \bar{p}. \tag{4.120}$$

(b) *The expected profit of a generator is obtained by*

$$\pi_d^*(c) = k[\bar{p} - c]. \tag{4.121}$$

(c) *The expected contribution margin per capacity unit is given by*

$$\pi_d^{u^*}(c) = \bar{p} - c. \tag{4.122}$$

(d) *The expected average price is given by*

$$E[p_d] = \bar{p}. \tag{4.123}$$

Proof.

(a) - **(d)** are easily checked with Theorem 4.10 (a) and (d) - (f). □

4.4 Uniform-Price Auction

4.4.1 Case A: A Single Generator Serves the Demand

The uniform price is set by the marginal bid. It is the bid of the winner if demand does not exceeds the capacity of a single generator. Hence, the winner is aware that he sets the price. The competitive situation is consequently the same as for the discriminatory auction where every generator gets his own bid paid. It is not surprising, that results from Theorems 4.2, 4.4, 4.3 (b), and 4.5 hold true for this auction type.

Theorem 4.11. *Suppose $d \in (0,k]$. The unique Bayes-Nash equilibrium in symmetric and strictly increasing bidding strategies is given by (b_u^*, b_u^*) with the following bid function*

$$b_u^*(c) = \left[\bar{p} - \bar{c} + \int_c^{\bar{c}} \exp\left(\int_\tau^{\bar{c}} \frac{f(t|t)}{1-F(t|t)}\,dt\right) d\tau\right] \exp\left(-\int_c^{\bar{c}} \frac{f(t|t)}{1-F(t|t)}\,dt\right) + c. \tag{4.124}$$

If marginal costs are independently drawn, the expression reduces to

$$b_u^{*i}(c) = \begin{cases} c + \dfrac{\bar{c} - c - \int_c^{\bar{c}} F(\tau)\,d\tau}{1-F(c)} & : c \in [\underline{c}, \bar{c}) \\ \bar{p} & : c = \bar{c}. \end{cases} \tag{4.125}$$

Proof.

Demand is sufficient for only one generator i.e. the winner of the auction takes it all. Taking into account the case of a standoff, the expected profit of generator i, with $j \neq i$, is given by

$$\pi_i(b_i, c_i) = E\left[d[b_j - c_i]1_{\{b_i<b_j\}} \,\middle|\, C_i = c_i\right].$$

The profit can also be written as follows, using definitions (4.1) and (4.2) of the discriminatory auction

$$\pi_i(b_i, c_i) = kE\left[\alpha[b_i - c_i]1_{\{b_i<b_j\}} + \beta[b_i - c_i]1_{\{b_i>b_j\}} \,\middle|\, C_i = c_i\right]. \tag{4.126}$$

It is not surprising that this expression equals the profit function (4.22) from the discriminatory auction. In this environment, the winning generator sets the uniform price and therefore the price he is paid. The decision problem is the same, as the expected profit shows. Consequently, all results of Theorem 4.2 apply for this case. □

Theorem 4.12. *Suppose $d \in (0,k]$.*

(a) *The auction leads to production efficiency. Production costs are the lowest possible.*

(b) *The expected production costs per electricity unit are given by*

$$E[c_u] = 2\int_{\underline{c}}^{\bar{c}} cF(c)\left[f(c|c) + F_i(c|c)\right]dc - 2\int_{\underline{c}}^{\bar{c}} F(c)\left[1 - F(c|c)\right]dc. \tag{4.127}$$

The expression for independently distributed marginal costs reduces to

$$E\left[c_u^i\right] = \bar{c} + \int_{\underline{c}}^{\bar{c}} F(c)^2\,dc - 2\int_{\underline{c}}^{\bar{c}} F(c)\,dc. \tag{4.128}$$

Proof.

(a) Optimal bid function (4.124) increases with marginal costs due to (4.32) and represents bids, which are accepted bids according to the auction rules. Hence, Theorem 4.1 (a) applies.

(b) Based on the same reason given in (a), (4.4) can be used. □

Theorem 4.13. *Suppose $d \in (0,k]$.*

(a) *The expected profit for one generator is obtained by*

$$\pi_u^*(c) = d[1 - F(c|c)]\left[\bar{p} - \bar{c} + \int_{c}^{\bar{c}} \exp\left(\int_{\tau}^{\bar{c}} \frac{f(t|t)}{1 - F(t|t)}\,dt\right)d\tau\right] \times \exp\left(-\int_{c}^{\bar{c}} \frac{f(t|t)}{1 - F(t|t)}\,dt\right). \tag{4.129}$$

For independently distributed marginal costs, the expression reduces to

$$\pi_u^{*i}(c) = d\left[\bar{c} - c - \int_c^{\bar{c}} F(\tau)\,d\tau\right]. \tag{4.130}$$

(b) *The expected contribution margin per capacity unit is given by*

$$\pi_u^{u^*}(c) = \frac{d}{k}[1 - F(c|c)]\left[\bar{p} - \bar{c} + \int_c^{\bar{c}} \exp\left(\int_\tau^{\bar{c}} \frac{f(t|t)}{1 - F(t|t)}\,dt\right)d\tau\right]$$
$$\times \exp\left(-\int_c^{\bar{c}} \frac{f(t|t)}{1 - F(t|t)}\,dt\right). \tag{4.131}$$

If marginal costs are independently drawn, the expression reduces to

$$\pi_u^{u*i}(c) = \frac{d}{k}\left[\bar{c} - c - \int_c^{\bar{c}} F(\tau)\,d\tau\right]. \tag{4.132}$$

Proof.

It was shown in the proof of Theorem 4.11 that the profit calculation and the decision problem of a generator here is the same as in the discriminatory auction. Hence, all other results of the discriminatory auction for the same demand range apply for this auction.

(a) The expected profit is given by (4.60). The expression for independently distributed marginal costs is obtained by (4.61).

(b) The expected contribution margin for each electricity unit is given by (4.62). For independently distributed costs, it is obtained by (4.63). □

Theorem 4.14. *Suppose $d \in (0,k]$. The expected uniform price is given by*

$$E[p_u] = 2\int_{\underline{c}}^{\bar{c}} F(c)F_i(c|c)\left[\bar{p} - \bar{c} + \int_c^{\bar{c}} \exp\left(\int_\tau^{\bar{c}} \frac{f(t|t)}{1 - F(t|t)}\,dt\right)d\tau\right]$$
$$\times \exp\left(-\int_c^{\bar{c}} \frac{f(t|t)}{1 - F(t|t)}\,dt\right)dc$$
$$+2\int_{\underline{c}}^{\bar{c}} cF(c)\left[f(c|c) + F_i(c|c)\right]dc. \tag{4.133}$$

For independently distributed marginal costs, the expression reduces to

$$E\left[p_u^i\right] = \bar{c} - \int_{\underline{c}}^{\bar{c}} F(c)^2 \, dc. \tag{4.134}$$

Proof.

Theorem 4.5 applies here due to the same argument given in the proof of Theorem 4.13. The expected uniform price equals the expected average price in the discriminatory auction because there exists only one winner. The price is therefore given by (4.73). The expression is for independently distributed marginal costs obtained by (4.74). □

Example 4.9. *Suppose the probability distribution given by Definition 3.5.*

(a) *The optimal bid function for independently distributed marginal costs is given by*

$$b_u^{*i}(c) = \begin{cases} \dfrac{1+c}{2} & : c \in [0,1) \\ \bar{p} & : c = 1. \end{cases} \tag{4.135}$$

For strong affiliation with $a > 0$, the optimal bid is obtained by

$$b_u^*(c) = \left\{ \bar{p} - 1 + \int_c^1 \exp\left(\int_\tau^1 \frac{2\left[1+at^2\right]}{2+at-[2+at^2]\,t} \, dt \right) d\tau \right\} \times \exp\left(-\int_c^1 \frac{2\left[1+at^2\right]}{2+at-[2+at^2]\,t} \, dt \right) + c. \tag{4.136}$$

(b) *The expected profit of a generator is given by*

$$\pi_u^{*i}(c) = d \, \frac{[1-c]^2}{2}. \tag{4.137}$$

Given strong affiliation with $a > 0$, the profit is obtained by

$$\pi_u^*(c) = d\left[1 - \frac{2+ac^2}{2+ac}\,c\right] \exp\left(-\int_c^1 \frac{2\left[1+at^2\right]}{2+at-[2+at^2]\,t} \, dt \right) \times \left\{ \bar{p} - 1 + \int_c^1 \exp\left(\int_\tau^1 \frac{2\left[1+at^2\right]}{2+at-[2+at^2]\,t} \, dt \right) d\tau \right\}. \tag{4.138}$$

(c) *For independently distributed marginal costs, the expected contribution margin per capacity unit is given by*

$$\pi_u^{u^{*i}}(c) = \frac{d}{k}\frac{[1-c]^2}{2}. \tag{4.139}$$

The contribution margin for strong affiliated costs with $a > 0$ is obtained by

$$\pi_u^{u^*}(c) = \frac{d}{k}\left[1-\frac{2+ac^2}{2+ac}c\right]\exp\left(-\int_c^1 \frac{2\left[1+at^2\right]}{2+at-[2+at^2]t}dt\right)$$
$$\times\left\{\bar{p}-1+\int_c^1 \exp\left(\int_\tau^1 \frac{2\left[1+at^2\right]}{2+at-[2+at^2]t}dt\right)d\tau\right\}. \tag{4.140}$$

(d) *The expected uniform price is given by*

$$E\left[p_u^i\right] = \frac{2}{3}. \tag{4.141}$$

The same expression for strong affiliation is obtained by

$$E\left[p_u\right] = \frac{4}{5}-\frac{2}{4+a}\left[\frac{1}{10}+\frac{2}{3a}-\frac{4}{a^2}-\frac{16}{a^3}+16\frac{2+a}{a^4}\ln\left(1+\frac{a}{2}\right)\right]$$
$$-\frac{4a}{4+a}\int_0^1 \frac{4+ac}{[2+ac]^2}[1-c]c^2\left\{\bar{p}-1\right.$$
$$\left.+\int_c^1 \exp\left(\int_\tau^1 \frac{2\left[1+at^2\right]}{2+at-[2+at^2]t}dt\right)d\tau\right\}$$
$$\times\exp\left(-\int_c^1 \frac{2\left[1+at^2\right]}{2+at-[2+at^2]t}dt\right)dc. \tag{4.142}$$

Proof.

(a) Results of Theorem 4.11 are the same as of Theorem 4.2. Hence, results of the derived Example 4.2 are valid here.

(b) Because Theorem 4.13 (a) has results identical to Theorem 4.4 (a), derived examples must have the same results, see Example 4.3 (a).

(c) Results of Example 4.3 (b) apply here because Theorems 4.13 (b) and 4.4 (b) have identical results.

(d) Due to identical results of Theorems 4.14 and 4.5, derived examples lead to the same results, which are given by Example 4.4. □

4.4.2 Case B: Both Generators Are Necessary

The decision problem of case B, with $d \in (k, 2k)$, differs from that in the discriminatory auction. According to the rules, the bid of the last selected bidder determines the uniform price. This eliminates the objective of high bids as a source of profit for the winner. He has only one aim left, which is maximizing the chance of winning. Hence, the bidding should be more aggressive.

The increased aggressiveness with low bids vanishes with higher demand. This results from the effect for the loser. The higher the demand electricity the higher the assured utilized production capacity. This relaxes the fear of lost opportunities. In turn, it allows a shift of focus from increasing the chance of winning at low demand to raising the price paid at a higher demand.

Based on the incentives for a bidder discussed above, the auctioneer can see lower as well as higher bids in this demand range compared with case A ($d \in (0, k]$). Nevertheless, all bids are higher than marginal costs. Because no generator is immune to losing an auction or sharing demand in a standoff, at least a small fraction of the bid refers to the incentive of high bids. Hence, bid shading should occur.

Theorem 4.15. *Suppose $d \in (k, 2k)$. The unique Bayes–Nash equilibrium in symmetric and strictly increasing bidding strategies is given by (b_u^*, b_u^*) with the following bid function*

$$b_u^*(c) = \left[\bar{p} - \bar{c} + \int_c^{\bar{c}} \exp\left(\frac{2k-d}{d-k} \int_\tau^{\bar{c}} \frac{f(t|t)}{F(t|t)}\, dt\right) d\tau\right] \times \exp\left((-)\frac{2k-d}{d-k} \int_c^{\bar{c}} \frac{f(t|t)}{F(t|t)}\, dt\right) + c. \tag{4.143}$$

under the condition that

$$\frac{1}{b_u^*(c) - c} > \frac{f_i(c|c)}{f(c|c)} - \frac{F_i(c|c)}{F(c|c)} \tag{4.144}$$

holds true for all marginal costs, or, which is equivalent, under the condition that $\bar{p}_u$ *exists and that*

$$\bar{p} < \bar{p}_u. \tag{4.145}$$

$\bar{p}_u$ *is the upper limit of the price cap and is obtained by*

$$\bar{p}_u = \bar{c} + \min\left(\left[\frac{1}{\frac{f_i(c|c)}{f(c|c)} - \frac{F_i(c|c)}{F(c|c)}} - \int_c^{\bar{c}} \exp\left((-)\frac{2k-d}{d-k}\int_c^{\tau} \frac{f(t|t)}{F(t|t)}\,dt\right) d\tau\right]\right.$$
$$\left.\times \exp\left(\frac{2k-d}{d-k}\int_c^{\bar{c}} \frac{f(t|t)}{F(t|t)}\,dt\right)\right), \tag{4.146}$$

for all marginal costs with

$$0 < \frac{f_i(c|c)}{f(c|c)} - \frac{F_i(c|c)}{F(c|c)}. \tag{4.147}$$

There exists no $\bar{p}_u$ *if no marginal costs can be found that satisfy (4.147).*

If marginal costs are independently drawn, the condition is always satisfied and the expression of the optimal bid function reduces to

$$b_u^{*i}(c) = [\bar{p} - \bar{c}]F(c)^{\frac{2k-d}{d-k}} + \int_c^{\bar{c}} \left[\frac{F(c)}{F(\tau)}\right]^{\frac{2k-d}{d-k}} d\tau + c. \tag{4.148}$$

Proof.

The proof is done in four steps. First, the bid function is found which maximizes the expected profit. Second, it is shown that the auction really takes place. Third, it is proved that a Bayes–Nash equilibrium is obtained by the bid function, and finally that only one Bayes–Nash equilibrium exists. Due to the symmetry of the generators, the proof is reduced to the analysis of generator i.

Step 1: Bid Function

Both generators are needed to serve demand in this environment. Therefore, all the cases of auction outcomes such as winning, losing, and a standoff have to be incorporated in the calculation of the expected profit. The formula for generator i with $j \neq i$, using (4.2), is given by

$$\begin{aligned}
\pi_i(b_i, c_i) &= E\left[k[b_j - c_i]1_{\{b_i<b_j\}} + [d-k][b_i - c_i]1_{\{b_i>b_j\}} \,\middle|\, C_i = c_i\right] \\
&= k\int_{b^{-1}(b_i)}^{\bar{c}} [b(c_j) - c_i]f(c_j|c_i)\,dc_j + k\beta[b_i - c_i]\int_{\underline{c}}^{b^{-1}(b_i)} f(c_j|c_i)\,dc_j \\
&= k\Bigg\{\beta[b_i - c_i]F\left(b^{-1}(b_i)\,|\,c_i\right) \\
&\quad + \int_{b^{-1}(b_i)}^{\bar{c}} [b(c_j) - c_i]f(c_j|c_i)\,dc_j\Bigg\}.
\end{aligned} \tag{4.149}$$

The first derivative of the expected profit is given by, taking into account (4.3)

$$\begin{aligned}
\frac{\partial \pi_i(b_i, c_i)}{\partial b_i} &= k\left\{\beta F\left(b^{-1}(b_i)\,|\,c_i\right) - [1-\beta][b_i - c_i]\frac{f\left(b^{-1}(b_i)\,|\,c_i\right)}{b\left(b^{-1}(b_i)\right)'}\right\} \\
&= k\left\{\beta F\left(b^{-1}(b_i)\,|\,c_i\right) - \beta_f[b_i - c_i]\frac{f\left(b^{-1}(b_i)\,|\,c_i\right)}{b\left(b^{-1}(b_i)\right)'}\right\}.
\end{aligned} \tag{4.150}$$

The necessary condition of an optimum $\frac{\partial \pi_i(b_i,c_i)}{\partial b_i} \stackrel{!}{=} 0$ is used to get the optimal bid function. The condition is equivalent to the following, see (4.150)

$$0 = b\left(b^{-1}(b_i)\right)'\beta F\left(b^{-1}(b_i)\,|\,c_i\right) - [b_i - c_i]\beta_f f\left(b^{-1}(b_i)\,|\,c_i\right).$$

$b_i = b(c_i)$ has to be valid for a symmetric Bayes–Nash equilibrium. Hence, the profit maximum is given if the following holds true

$$b(c)' - b(c)\frac{\beta_f f(c|c)}{\beta F(c|c)} = (-)c\frac{\beta_f f(c|c)}{\beta F(c|c)}. \tag{4.151}$$

A solution for the linear non-homogeneous differential equation is[4]

[4] For a general solution of linear first-order differential equations see Sydsæter, Strøm, and Berck (1999), p. 62.

$$b(c) = b_0 \mathrm{e}^{\frac{\beta_f}{\beta}\int_\theta^c \frac{f(t|t)}{F(t|t)}dt} - \frac{\beta_f}{\beta}\int_\theta^c \tau \frac{f(\tau|\tau)}{F(\tau|\tau)} \mathrm{e}^{\frac{\beta_f}{\beta}\int_\tau^c \frac{f(t|t)}{F(t|t)}dt} d\tau$$

$$= b_0 \mathrm{e}^{\frac{\beta_f}{\beta}\int_\theta^c \frac{f(t|t)}{F(t|t)}dt} + \tau \mathrm{e}^{\frac{\beta_f}{\beta}\int_\tau^c \frac{f(t|t)}{F(t|t)}dt}\Big|_\theta^c - \int_\theta^c \mathrm{e}^{\frac{\beta_f}{\beta}\int_\tau^c \frac{f(t|t)}{F(t|t)}dt} d\tau$$

$$= c + [b_0 - \theta]\mathrm{e}^{\frac{\beta_f}{\beta}\int_\theta^c \frac{f(t|t)}{F(t|t)}dt} - \int_\theta^c \mathrm{e}^{\frac{\beta_f}{\beta}\int_\tau^c \frac{f(t|t)}{F(t|t)}dt} d\tau. \tag{4.152}$$

The upper limit of the bid function is set by $\hat{b}$, see (4.28)

$$b(\bar{c}) = \hat{b} \tag{4.153}$$

$$\Longleftrightarrow$$

$$\hat{b} = \bar{c} + [b_0 - \theta]\mathrm{e}^{\frac{\beta_f}{\beta}\int_\theta^{\bar{c}} \frac{f(t|t)}{F(t|t)}dt} - \int_\theta^{\bar{c}} \mathrm{e}^{\frac{\beta_f}{\beta}\int_\tau^{\bar{c}} \frac{f(t|t)}{F(t|t)}dt} d\tau$$

$$\Longleftrightarrow$$

$$b_0 = \theta + \left[\hat{b} - \bar{c} + \int_\theta^{\bar{c}} \mathrm{e}^{\frac{\beta_f}{\beta}\int_\tau^{\bar{c}} \frac{f(t|t)}{F(t|t)}dt} d\tau\right] \mathrm{e}^{(-)\frac{\beta_f}{\beta}\int_\theta^{\bar{c}} \frac{f(t|t)}{F(t|t)}dt}.$$

Substituting b_0, the bid function (4.152) is given by

$$b(c) = \left[\hat{b} - \bar{c} + \int_\theta^{\bar{c}} \mathrm{e}^{\frac{\beta_f}{\beta}\int_\tau^{\bar{c}} \frac{f(t|t)}{F(t|t)}dt} d\tau\right] \mathrm{e}^{(-)\frac{\beta_f}{\beta}\int_c^{\bar{c}} \frac{f(t|t)}{F(t|t)}dt} - \int_\theta^c \mathrm{e}^{\frac{\beta_f}{\beta}\int_\tau^c \frac{f(t|t)}{F(t|t)}dt} d\tau + c$$

$$= \left[\hat{b} - \bar{c}\right] \mathrm{e}^{(-)\frac{\beta_f}{\beta}\int_c^{\bar{c}} \frac{f(t|t)}{F(t|t)}dt} + \int_\theta^{\bar{c}} \mathrm{e}^{\frac{\beta_f}{\beta}\int_\tau^c \frac{f(t|t)}{F(t|t)}dt} d\tau - \int_\theta^c \mathrm{e}^{\frac{\beta_f}{\beta}\int_\tau^c \frac{f(t|t)}{F(t|t)}dt} d\tau + c$$

$$= \left[\hat{b} - \bar{c}\right] \mathrm{e}^{(-)\frac{\beta_f}{\beta}\int_c^{\bar{c}} \frac{f(t|t)}{F(t|t)}dt} + \int_c^{\bar{c}} \mathrm{e}^{\frac{\beta_f}{\beta}\int_\tau^c \frac{f(t|t)}{F(t|t)}dt} d\tau + c$$

$$= \left[\hat{b} - \bar{c} + \int_c^{\bar{c}} \mathrm{e}^{\frac{\beta_f}{\beta}\int_\tau^{\bar{c}} \frac{f(t|t)}{F(t|t)}dt} d\tau\right] \mathrm{e}^{(-)\frac{\beta_f}{\beta}\int_c^{\bar{c}} \frac{f(t|t)}{F(t|t)}dt} + c. \tag{4.154}$$

The first derivative of the bid function follows directly from (4.151)

$$\frac{\partial b(c)}{\partial c} = \frac{\beta_f f(c|c)}{\beta F(c|c)} [b(c) - c]. \tag{4.155}$$

Based on (4.155), the second derivative of the bid function is obtained by

$$\begin{aligned}
\frac{\partial^2 b(c)}{\partial c^2} &= \frac{\beta_f}{\beta} \left\{ [b(c)' - 1] \frac{f(c|c)}{F(c|c)} + [b(c) - c] \frac{f_j(c|c) + f_i(c|c)}{F(c|c)} \right. \\
&\qquad \left. - [b(c) - c] f(c|c) \frac{f(c|c) + F_i(c|c)}{F(c|c)^2} \right\} \\
&= [b(c)' - 1] \frac{\beta_f}{\beta} \frac{f(c|c)}{F(c|c)} + b(c)' \frac{f_j(c|c) + f_i(c|c)}{f(c|c)} \\
&\qquad - b(c)' \frac{f(c|c) + F_i(c|c)}{F(c|c)}.
\end{aligned} \tag{4.156}$$

It was presumed that the bid function is strictly increasing. This assumption was correct only if following condition holds true

$$\frac{\partial b(c)}{\partial c} > 0, \quad \forall c \in [\underline{c}, \bar{c}]. \tag{4.157}$$

Because $\frac{f(c|c)}{F(c|c)}$ is always positive, the condition is equivalent to, see (4.155)

$$b(c) > c. \tag{4.158}$$

According to bid function (4.154), this inequation is valid because $\hat{b} > \bar{c}$, see (4.28). Hence, the bid function is strictly increasing as presumed.

The bid function leads to the profit maximum if the second derivative of the expected profit bidding according to $b(c)$ is negative, i.e.

$$\frac{\partial^2 \pi(b(c), c)}{\partial b(c)^2} < 0. \tag{4.159}$$

Using (4.150), the second derivative is determined by

$$\begin{aligned}
&\frac{\partial^2 \pi_i(b_i, c_i)}{\partial b_i^2} \\
&= k \left\{ \beta \frac{f\left(b^{-1}(b_i) | c_i\right)}{b\left(b^{-1}(b_i)\right)'} - \beta_f \frac{f\left(b^{-1}(b_i) | c_i\right)}{b\left(b^{-1}(b_i)\right)'} - \beta_f [b_i - c_i] \right. \\
&\qquad \left. \times \frac{f_j\left(b^{-1}(b_i) | c_i\right)}{\left[b\left(b^{-1}(b_i)\right)'\right]^2} + \beta_f [b_i - c_i] \frac{f\left(b^{-1}(b_i) | c_i\right) b\left(b^{-1}(b_i)\right)''}{\left[b\left(b^{-1}(b_i)\right)'\right]^3} \right\}
\end{aligned}$$

$$= k[\beta - \beta_f] \frac{f\left(b^{-1}(b_i) \,|\, c_i\right)}{b\left(b^{-1}(b_i)\right)'} - k\beta_f \frac{b_i - c_i}{\left[b\left(b^{-1}(b_i)\right)'\right]^3}$$
$$\times \left[f_j\left(b^{-1}(b_i) \,|\, c_i\right) b\left(b^{-1}(b_i)\right)' - f\left(b^{-1}(b_i) \,|\, c_i\right) b\left(b^{-1}(b_i)\right)''\right].$$

The profit derivative is evaluated at $b_i = b(c)$ because generators bid according to bid function (4.154). Hence, the profit derivative, using bid derivatives (4.155) and (4.156), is obtained by

$$\frac{\partial^2 \pi (b(c), c)}{\partial b(c)^2}$$
$$= k[\beta - \beta_f] \frac{f(c|c)}{b(c)'} - k\beta_f [b(c) - c] \frac{f_j(c|c) b(c)' - f(c|c) b(c)''}{[b(c)']^3}$$
$$= k[\beta - \beta_f] \frac{f(c|c)}{b(c)'} - k\beta \frac{f_j(c|c) F(c|c)}{f(c|c) b(c)'} + k\beta \frac{F(c|c) b(c)''}{[b(c)']^2}$$
$$= k[\beta - \beta_f] \frac{f(c|c)}{b(c)'} - k\beta \frac{f_j(c|c) F(c|c)}{f(c|c) b(c)'} + k\beta_f \frac{b(c)' - 1}{[b(c)']^2} f(c|c)$$
$$+ k\beta \frac{f_j(c|c) + f_i(c|c)}{f(c|c) b(c)'} F(c|c) - k\beta \frac{f(c|c) + F_i(c|c)}{b(c)'}$$
$$= k[\beta - \beta_f] \frac{f(c|c)}{b(c)'} + k\beta_f \frac{b(c)' - 1}{[b(c)']^2} f(c|c) + k\beta \frac{f_i(c|c) F(c|c)}{f(c|c) b(c)'}$$
$$- k\beta \frac{f(c|c) + F_i(c|c)}{b(c)'}$$
$$= k \left\{ \beta \frac{f_i(c|c) F(c|c) - f(c|c) F_i(c|c)}{f(c|c) b(c)'} - \beta_f \frac{f(c|c)}{[b(c)']^2} \right\}$$
$$= k \frac{\beta_f}{[b(c)']^2} \left\{ [b(c) - c] \left[f_i(c|c) - \frac{f(c|c) F_i(c|c)}{F(c|c)} \right] - f(c|c) \right\}. \quad (4.160)$$

The second derivative is negative for all generators if the following condition holds true

$$[b(c)-c]\left[f_i(c|c)-\frac{f(c|c)F_i(c|c)}{F(c|c)}\right]-f(c|c)<0$$

$$\Longleftrightarrow$$

$$\frac{f(c|c)}{b(c)-c}>f_i(c|c)-\frac{f(c|c)F_i(c|c)}{F(c|c)}$$

$$\Longleftrightarrow$$

$$\frac{1}{b(c)-c}>\frac{f_i(c|c)}{f(c|c)}-\frac{F_i(c|c)}{F(c|c)}. \tag{4.161}$$

This condition must be valid only because β never reaches one, see (4.2). The other reason is that bids are always greater than marginal costs according to (4.158) which implies that the first derivative of the bid function is not zero, see (4.157).

Given independently distributed marginal costs, the condition, using (3.11), (3.12), (3.13), and (3.14), can be rewritten to

$$\frac{1}{b^i(c)-c}>\frac{f_i(c|c)}{f(c|c)}-\frac{F_i(c|c)}{F(c|c)}$$
$$>0.$$

This inequation holds always true because $b^i(c)>c$ due to (4.158). Hence, condition (4.161) is only necessary for other distributions and holds true due to (4.144). Another expression of this condition is given at the end of the proof to use the found optimal bidding strategy, see (4.171).

Step 2: Participation of Generators

The auction really takes place if all generators have an incentive to participate. That is equivalent to

$$\pi(b(c),c)>0. \tag{4.162}$$

Expected profit (4.149), using bid function (4.154), is determined by

$$\pi(b(c),c)=k\left\{\beta[b(c)-c]F(c|c)+\int_c^{\bar{c}}[b(\tau)-c]f(\tau|c)\,d\tau\right\}. \tag{4.163}$$

The expression is positive because at first bids are higher than marginal costs, see (4.154). Additionally, (4.2) states that β never becomes zero and the probability that a generator has a specified marginal costs is not zero. Hence, the inequation (4.162) is holds true and all generators participate in the auction.

Step 3: Existence of a Bayes–Nash Equilibrium

The existence of a Bayes–Nash equilibrium is proved by showing that generator i bids according to the bid function $b(c_i)$ while the other generators keep using the same bid function. Suppose a generator wants to bid differently from bid function (4.154) but does not go below the lowest bid $b(\underline{c})$ or above the highest bid $b(\bar{c})$. In this case, the profit maximizing strategy is not to deviate from (4.154) because it was shown earlier that this bid function yields the highest profit.

Undercutting the lowest bid in equilibrium, $b(\underline{c})$, increases the profit if the following holds true for $\varepsilon > 0$ and $b(\underline{c}) - \varepsilon \geq c$

$$\frac{\partial \pi (b(\underline{c}) - \varepsilon, c)}{\partial b(\underline{c}) - \varepsilon} < 0. \tag{4.164}$$

The expected profit is determined by

$$\pi (b(\underline{c}) - \varepsilon, c) = d [b(\underline{c}) - \varepsilon - c],$$

because winning the auction is sure in the case of undercutting. The first derivative of the profit is then

$$\frac{\partial \pi (b(\underline{c}) - \varepsilon, c)}{\partial b(\underline{c}) - \varepsilon} = d.$$

It is easy to see that (4.164) does not hold true because $d > 0$. Therefore, undercutting is not a profit increasing strategy. Making extra profit by overbidding the highest optimal bid is only possible for $b(\bar{c}) + \varepsilon \leq \bar{p}$ and $\varepsilon > 0$ if the following holds true

$$\pi (b(\bar{c}) + \varepsilon, c) - \pi (b(c), c) > 0. \tag{4.165}$$

Overbidding the highest bid and earning money is only possible for $\hat{b} < \bar{p}$. In this case and if no other generator changes his behavior, the profit of overbidding is

$$\pi (b(\bar{c}) + \varepsilon, c) = k\beta [b(\bar{c}) + \varepsilon - c] = k\beta \left[\hat{b} - c + \varepsilon\right]. \tag{4.166}$$

Using (4.166) and (4.163), the extra profit of overbidding for $b(\bar{c})+\varepsilon \leq \bar{p}$ and $\varepsilon > 0$ is obtained by

$$\begin{aligned}&\pi\left(b(\bar{c})+\varepsilon,c\right)-\pi\left(b\left(c\right),c\right)\\&\quad=k\left\{\beta\left[\hat{b}-c+\varepsilon\right]-\beta\left[b\left(c\right)-c\right]F\left(c|c\right)-\int_{c}^{\bar{c}}\left[b(\tau)-c\right]f\left(\tau|c\right)d\tau\right\}.\end{aligned} \tag{4.167}$$

Extra profit increases with higher ε because

$$\frac{\partial}{\partial\varepsilon}\left[\pi\left(b(\bar{c})+\varepsilon,c\right)-\pi\left(b\left(c\right),c\right)\right]=k\beta,$$

is positive. To avoid this outcome and therefore the deviation from the bid function, ε has to be set to zero. That is done if the following holds true, see (4.153)

$$\begin{aligned}&b(\bar{c})=\bar{p}\\&\Longleftrightarrow\\&\hat{b}\quad=\bar{p}.\end{aligned}$$

Hence, the bid function, using (4.154), is determined by

$$b\left(c\right)=\left[\bar{p}-\bar{c}+\int_{c}^{\bar{c}}\mathrm{e}^{\frac{\beta_f}{\beta}\int_{\tau}^{\bar{c}}\frac{f(t|t)}{F(t|t)}dt}d\tau\right]\mathrm{e}^{(-)\frac{\beta_f}{\beta}\int_{c}^{\bar{c}}\frac{f(t|t)}{F(t|t)}dt}+c. \tag{4.168}$$

Step 4: Uniqueness of the Bayes–Nash Equilibrium

It was shown that bid function (4.168) is under condition (4.161) the only function that maximizes the expected profit. There is no remaining free variable for the generator to choose. Hence, (4.168) determines the bidding strategy for each generator in a Bayes–Nash equilibrium that is unique.

Substituting (4.2) and (4.3) for β and β_f respectively, the optimal bid function is obtained by

$$b_u^*\left(c\right)=\left[\bar{p}-\bar{c}+\int_{c}^{\bar{c}}\mathrm{e}^{\frac{2k-d}{d-k}\int_{\tau}^{\bar{c}}\frac{f(t|t)}{F(t|t)}dt}d\tau\right]\mathrm{e}^{(-)\frac{2k-d}{d-k}\int_{c}^{\bar{c}}\frac{f(t|t)}{F(t|t)}dt}+c. \tag{4.169}$$

The condition stated in (4.161) can be rewritten to the following expression for all marginal costs, which satisfy

$$0<\frac{f_i\left(c|c\right)}{f\left(c|c\right)}-\frac{F_i\left(c|c\right)}{F\left(c|c\right)}, \tag{4.170}$$

because these marginal costs generate positive values at the right term of (4.161), which makes it necessary to prove the condition because the left term of (4.161) is always positive due to (4.158). The condition, using bid function (4.169), is given by

$$\frac{1}{\frac{f_i(c|c)}{f(c|c)} - \frac{F_i(c|c)}{F(c|c)}} > b_u^*(c) - c$$

$$> \left[\bar{p} - \bar{c} + \int_c^{\bar{c}} e^{\frac{2k-d}{d-k} \int_\tau^{\bar{c}} \frac{f(t|t)}{F(t|t)} dt} d\tau\right] e^{(-)\frac{2k-d}{d-k} \int_c^{\bar{c}} \frac{f(t|t)}{F(t|t)} dt}$$

$$\Longleftrightarrow$$

$$\frac{e^{\frac{2k-d}{d-k} \int_c^{\bar{c}} \frac{f(t|t)}{F(t|t)} dt}}{\frac{f_i(c|c)}{f(c|c)} - \frac{F_i(c|c)}{F(c|c)}} > \bar{p} - \bar{c} + \int_c^{\bar{c}} e^{\frac{2k-d}{d-k} \int_\tau^{\bar{c}} \frac{f(t|t)}{F(t|t)} dt} d\tau$$

$$\Longleftrightarrow$$

$$\bar{p} < \min\left(e^{\frac{2k-d}{d-k} \int_c^{\bar{c}} \frac{f(t|t)}{F(t|t)} dt} \left[\frac{1}{\frac{f_i(c|c)}{f(c|c)} - \frac{F_i(c|c)}{F(c|c)}} - \int_c^{\bar{c}} e^{(-)\frac{2k-d}{d-k} \int_c^{\tau} \frac{f(t|t)}{F(t|t)} dt} d\tau\right]\right)$$
$$+\bar{c}. \tag{4.171}$$

For independently distributed marginal costs, the expression, using (3.11), (3.12), and (3.3), is given by

$$b_u^{*i}(c) = \left[\bar{p} - \bar{c} + \int_c^{\bar{c}} e^{\frac{2k-d}{d-k} \int_\tau^{\bar{c}} \frac{f(t)}{F(t)} dt} d\tau\right] e^{(-)\frac{2k-d}{d-k} \int_c^{\bar{c}} \frac{f(t)}{F(t)} dt} + c$$

$$= \left\{\bar{p} - \bar{c} + \int_c^{\bar{c}} \left[\frac{F(\bar{c})}{F(\tau)}\right]^{\frac{2k-d}{d-k}} d\tau\right\} \left[\frac{F(c)}{F(\bar{c})}\right]^{\frac{2k-d}{d-k}} + c$$

$$= [\bar{p} - \bar{c}] F(c)^{\frac{2k-d}{d-k}} + \int_c^{\bar{c}} \left[\frac{F(c)}{F(\tau)}\right]^{\frac{2k-d}{d-k}} d\tau + c.$$

□

Although the bidding strategy in the uniform-price auction differs from those in the discriminatory auction for the same demand range, there exists an upper limit of the price cap called $\bar{p}_u$. It must not be reached if $\bar{p}_u$ exists otherwise some generators will deviate for the stated bid function (4.143). Hence, the auctioneer must choose a price cap $\bar{p}$ below that value if it exists to ensure that all generators bid according to (4.143).

Corollary 4.7. *Suppose $d \in (k, 2k)$.*

(a) *Bid shading is positive, i.e. the optimal bid of a generator is greater than his marginal costs.*
(b) *The optimal bid function increases strictly with marginal costs.*
(c) *The optimal bid function increases strictly with demand for all marginal costs but $\bar{c}$. The optimal bid is independent of demand for $\bar{c}$.*
(d) *Generators bid their marginal costs for a demand close to the production capacity of a single generator k. The exception is the generator with the highest marginal cost $\bar{c}$. He bids the price cap $\bar{p}$.*
(e) *Optimal bids converge to price cap $\bar{p}$ when demand comes close to $2k$, the production capacity of the whole market.*

Proof.

(a) Optimal bids are greater than marginal costs, see (4.158). Hence, the difference between optimal bid minus marginal costs is greater than zero.
(b) A condition of the proof of Theorem 4.15 was that the optimal bid function is strictly increasing with marginal costs. It was shown there that this condition holds true, see (4.157).
(c) The first derivative of the bid function with respect to demand, using (4.143), is given by

$$\begin{aligned}\frac{\partial b_u^*(c)}{\partial d} &= k\,\frac{\bar{p}-\bar{c}}{[d-k]^2}\,\mathrm{e}^{\frac{2k-d}{d-k}\int_c^{\bar{c}}\frac{(-)f(t|t)}{F(t|t)}\,dt}\int_c^{\bar{c}}\frac{f(t|t)}{F(t|t)}\,dt \\ &\quad+\frac{k}{[d-k]^2}\int_c^{\bar{c}}\mathrm{e}^{\frac{2k-d}{d-k}\int_c^{\tau}\frac{(-)f(t|t)}{F(t|t)}\,dt}\int_c^{\tau}\frac{f(t|t)}{F(t|t)}\,dt\,d\tau \\ &= \frac{k}{[d-k]^2}\left\{[\bar{p}-\bar{c}]\mathrm{e}^{\frac{2k-d}{d-k}\int_c^{\bar{c}}\frac{(-)f(t|t)}{F(t|t)}\,dt}\int_c^{\bar{c}}\frac{f(t|t)}{F(t|t)}\,dt\right. \\ &\quad\left.+\int_c^{\bar{c}}\mathrm{e}^{\frac{2k-d}{d-k}\int_c^{\tau}\frac{(-)f(t|t)}{F(t|t)}\,dt}\int_c^{\tau}\frac{f(t|t)}{F(t|t)}\,dt\,d\tau\right\}. \end{aligned} \tag{4.172}$$

The derivative is positive if the following holds true, because $d > k$

$$0 < [\bar{p}-\bar{c}]\mathrm{e}^{\frac{2k-d}{d-k}\int\limits_{c}^{\bar{c}}\frac{(-)f(t|t)}{F(t|t)}dt}\int\limits_{c}^{\bar{c}}\frac{f(t|t)}{F(t|t)}dt$$

$$+\int\limits_{c}^{\bar{c}}\mathrm{e}^{\frac{2k-d}{d-k}\int\limits_{c}^{\tau}\frac{(-)f(t|t)}{F(t|t)}dt}\int\limits_{c}^{\tau}\frac{f(t|t)}{F(t|t)}dt\,d\tau.$$

The right term is zero for $\bar{c}$ because both summands have integrals as factors, which become zero for the highest marginal cost $\bar{c}$. Both summands of the right term are positive for all other marginal costs because $\bar{p} > \bar{c}$, see Assumption 3.12, and $f(c|c) > 0\ \forall\ c \in [\underline{c},\bar{c}]$, see Assumption 3.7. Hence, the optimal bid function strictly increases with demand for $c \in [\underline{c},\bar{c})$ due to a positive first derivative. The bid function for $\bar{c}$ is independent of demand because the slope equals zero.

(d) The optimal bid function (4.143) converges for $d \longrightarrow k_+$ and $c \in [\underline{c},\bar{c})$ to the following expression, taking into account that $f(c|c) > 0$ and $F(c|c) \geq 0$ due to (3.4) and (3.5) respectively

$$\lim_{d\to k_+} b_u^*(c) = \lim_{d\to k_+}\left[\bar{p}-\bar{c}+\int\limits_{c}^{\bar{c}}\mathrm{e}^{\frac{2k-d}{d-k}\int\limits_{\tau}^{\bar{c}}\frac{f(t|t)}{F(t|t)}dt}d\tau\right]\mathrm{e}^{\frac{2k-d}{d-k}\int\limits_{c}^{\bar{c}}\frac{(-)f(t|t)}{F(t|t)}dt}+c$$

$$= c+\lim_{\gamma\to 0_+}[\bar{p}-\bar{c}]\mathrm{e}^{\frac{(-)k}{\gamma}\int\limits_{c}^{\bar{c}}\frac{f(t|t)}{F(t|t)}dt}+\lim_{\gamma\to 0_+}\int\limits_{c}^{\bar{c}}\mathrm{e}^{\frac{(-)k}{\gamma}\int\limits_{c}^{\tau}\frac{f(t|t)}{F(t|t)}dt}d\tau$$

$$= c.$$

All integrals of the optimal bid are zero for $\bar{c}$. Hence, the optimal bid converges for $d \longrightarrow k_+$ and $\bar{c}$ to

$$\lim_{d\to k_+} b_u^*(\bar{c}) = \bar{p}.$$

(e) Optimal bids for $d \in (k,2k)$, see (4.143), converge for $d \longrightarrow 2k_-$ to

$$\lim_{d\to 2k_-} b_u^*(c) = \lim_{d\to 2k_-}\left[\bar{p}-\bar{c}+\int\limits_{c}^{\bar{c}}\mathrm{e}^{\frac{2k-d}{d-k}\int\limits_{\tau}^{\bar{c}}\frac{f(t|t)}{F(t|t)}dt}d\tau\right]\mathrm{e}^{\frac{2k-d}{d-k}\int\limits_{c}^{\bar{c}}\frac{(-)f(t|t)}{F(t|t)}dt}+c$$

$$= \bar{p}-\bar{c}+\int\limits_{c}^{\bar{c}}d\tau+c = \bar{p}.$$

□

Generators bid higher than their marginal costs, see Corollary 4.7 (a). But for a demand close to k, they bid only slightly higher than their marginal costs except the generator with marginal costs of $\bar{c}$, see Corollary 4.7 (d). They do so because the loser sets the price. Hence, generators speculate on a higher bid by their competitor. The advantage of the strategy is that the winner fully utilizes his production capacity. He accepts the possibility that the uniform price may lay only slightly above his marginal costs. It shows that making money by selling k electricity units instead of nearly nothing ($\lim_{d \to k_+}[d-k] = 0_+$) is more important than maximizing the price paid by bidding higher.

The ranking of the generators according to their bids leads to the same order as ranking them by their marginal costs. The reason is that optimal bids increase with marginal costs, see Corollary 4.7 (b). Optimal bids increase with demand, see Corollary 4.7 (c), because the assured served demand $d-k$ increases with higher demand. This lowers the pressure to win and allows bids to rise. The highest bids are reached for $d \to 2k_-$ due to Corollary 4.7 (e). They are almost close to $\bar{p}$.

Example 4.10.

Suppose the probability distribution given by Definition 3.5. The optimal bid function for independently distributed marginal costs is given by

$$b_u^{*i}(c) = \begin{cases} 0 & : d = \frac{3}{2}k,\ c = 0 \\ [\bar{p} - \ln(c)]c & : d = \frac{3}{2}k,\ c \in (0,1] \\ \left[\bar{p} - \frac{2k-d}{3k-2d}\right] c^{\frac{2k-d}{d-k}} + \frac{2k-d}{3k-2d} c & : d \neq \frac{3}{2}k. \end{cases} \tag{4.173}$$

For strong affiliated marginal costs with $a > 0$, the bid function is obtained by

$$b_u^*(c) = \left\{ \bar{p} - 1 + \int_c^1 \exp\left(\frac{2k-d}{d-k} \int_\tau^1 \frac{2[1+at^2]}{[2+at^2]t} dt \right) d\tau \right\} \times \exp\left(\frac{2k-d}{d-k} \int_c^1 \frac{(-2)[1+at^2]}{[2+at^2]t} dt \right) + c \tag{4.174}$$

if the following condition holds true for all marginal costs

$$\frac{1}{b_u^*(c) - c} > \frac{ac}{[1+ac^2][2+ac^2]}. \tag{4.175}$$

Equation (4.175) does not need to be checked if the following condition holds true for the price cap

$$\bar{p} < \begin{cases} a+4+\dfrac{2}{a} & : a < \frac{\sqrt{10}-2}{3} \\ \dfrac{\sqrt{10}+1}{3}+3\dfrac{\sqrt{10}}{\sqrt{10}-2} & : a \geq \frac{\sqrt{10}-2}{3}. \end{cases}$$

Proof.

Optimal bid function (4.143) is obtained by, taking into account (3.26) and (3.27)

$$b_u^*(c) = \left[\bar{p}-1+\int_c^1 \mathrm{e}^{\frac{2k-d}{d-k}\int_\tau^1 \frac{\frac{1+at^2}{1+\frac{a}{2}t}}{\frac{1+\frac{a}{2}t^2}{1+\frac{a}{2}t}t}dt} d\tau\right] \mathrm{e}^{(-)\frac{2k-d}{d-k}\int_c^1 \frac{\frac{1+at^2}{1+\frac{a}{2}t}}{\frac{1+\frac{a}{2}t^2}{1+\frac{a}{2}t}t}dt} + c$$

$$= \left\{\bar{p}-1+\int_c^1 \mathrm{e}^{\frac{2k-d}{d-k}\int_\tau^1 \frac{2[1+at^2]}{[2+at^2]t}dt} d\tau\right\} \mathrm{e}^{\frac{2k-d}{d-k}\int_c^1 \frac{(-2)[1+at^2]}{[2+at^2]t}dt} + c.$$

The expression for independently distributed marginal costs, using (4.148) and (3.25), is obtained by

$$b_u^{*i}(c) = \left[\bar{p}-1+\int_c^1 \tau^{(-)\frac{2k-d}{d-k}} d\tau\right] c^{\frac{2k-d}{d-k}} + c. \tag{4.176}$$

The bid function for $d = \frac{3}{2}k$ is given by

$$b_u^{*i}(c) = \left[\bar{p}+\int_c^1 \frac{1}{\tau} d\tau\right] c = \begin{cases} 0 & : c = 0 \\ [\bar{p}-\ln(c)]c & : c \in (0,1]. \end{cases}$$

For all other demands, (4.176) is obtained by

$$b_u^{*i}(c) = \left[\bar{p}-1-\frac{d-k}{3k-2d}\, \tau^{(-)\frac{3k-2d}{d-k}}\Big|_c^1\right] c^{\frac{2k-d}{d-k}} + c$$

$$= \left[\bar{p}-1-\frac{d-k}{3k-2d}+\frac{d-k}{3k-2d}c^{(-)\frac{3k-2d}{d-k}}\right] c^{\frac{2k-d}{d-k}} + c$$

$$= \left[\bar{p}-\frac{2k-d}{3k-2d}\right] c^{\frac{2k-d}{d-k}} + \frac{2k-d}{3k-2d}\, c.$$

The bid function for strong affiliated marginal costs is optimal if condition (4.161) holds true. Using (3.26), (3.27), (3.31), and (3.32), the expression is given by

$$\begin{aligned}\frac{1}{b_u^*(c)-c} &> \frac{\frac{a\left[c-\frac{1}{2}\right]}{\left[1+\frac{a}{2}c\right]^2}}{\frac{1+ac^2}{1+\frac{a}{2}c}} + \frac{\frac{a[1-c]c}{2\left[1+\frac{a}{2}c\right]^2}}{\frac{1+\frac{a}{2}c^2}{1+\frac{a}{2}c}c} \\ &> \frac{2a}{2+ac}\left[\frac{c-\frac{1}{2}}{1+ac^2}+\frac{1-c}{2+ac^2}\right] \\ &> \frac{ac}{[1+ac^2][2+ac^2]}.\end{aligned}$$

The condition always holds true without calculating the optimal bid if the following is satisfied. The reasons are that bid shading is never zero, see Corollary 4.4 (a), $a > 0$, and that the lowest left term is given by the highest possible contribution margin $\bar{p} - c$.

$$\begin{aligned}\frac{1}{\bar{p}-c} &> \frac{ac}{[1+ac^2][2+ac^2]} \\ &\Longleftrightarrow \\ ac[\bar{p}-c] &< [1+ac^2][2+ac^2] \\ &\Longleftrightarrow \\ ac\bar{p} &< a^2c^4+4ac^2+2.\end{aligned}$$

It is obvious that for $c = 0$ the inequation holds true. The expression for $c \in (0,1]$ can be rewritten to

$$\bar{p} < ac^3 + 4c + \frac{2}{ac}. \tag{4.177}$$

The lowest value of the right term concerning c is found for

$$\frac{\partial}{\partial c}\left[ac^3+4c+\frac{2}{ac}\right] \stackrel{!}{=} 0.$$

The derivative is given by

$$\frac{\partial}{\partial c}\left[ac^3+4c+\frac{2}{ac}\right] = 3ac^2+4-\frac{2}{ac^2}$$
$$= \frac{3a}{c^2}\left[c^4+\frac{4}{3a}c^2-\frac{2}{3a^2}\right].$$

Because $a > 0$, the derivative equals zero if the following holds true

$$0 = c^4+\frac{4}{3a}c^2-\frac{2}{3a^2}$$
$$= \left[c^2+\frac{2}{3a}\right]^2-\frac{10}{9a^2}$$
$$\Longleftrightarrow$$
$$c = \sqrt{\frac{\sqrt{10}-2}{3a}}. \tag{4.178}$$

It is obvious that the value is positive for $a > 0$. The result represents a minimum if the second derivative

$$\frac{\partial^2}{\partial c^2}\left[ac^3+4c+\frac{2}{ac}\right] = 6ac+\frac{4}{ac^3}$$
$$= c\left[6a+\frac{4}{ac^4}\right]$$

is positive evaluated at $c = \sqrt{\frac{\sqrt{10}-2}{3a}}$. The result is

$$\sqrt{\frac{\sqrt{10}-2}{3a}}\left[6a+\frac{36a}{[\sqrt{10}-2]^2}\right] = 6a\sqrt{\frac{\sqrt{10}-2}{3a}}\,\frac{6+[\sqrt{10}-2]^2}{[\sqrt{10}-2]^2}$$
$$= 24a\sqrt{\frac{\sqrt{10}-2}{3a}}\,\frac{5-\sqrt{10}}{[\sqrt{10}-2]^2}$$
$$= 8\,\frac{5-\sqrt{10}}{\sqrt{10}-2}\sqrt{\frac{3a}{\sqrt{10}-2}}.$$

It is easy to see that the term is positive for $a > 0$. Hence, (4.178) leads to a minimum of the right term of (4.177). Because the highest marginal costs is 1, (4.178) can only be used if the following holds true

$$1 \geq \sqrt{\frac{\sqrt{10}-2}{3a}}$$

$$\Longleftrightarrow$$

$$a \geq \frac{\sqrt{10}-2}{3}.$$

Given $a \geq \frac{\sqrt{10}-2}{3}$, (4.178) is used to rewrite condition (4.177) to

$$\begin{aligned}
\bar{p} &< \frac{1}{c^2}\left[ac^4 + 4c^2 + \frac{2}{a}\right] \\
&< \frac{3a}{\sqrt{10}-2}\left\{\frac{[\sqrt{10}-2]^2}{9a} + 4\frac{\sqrt{10}-2}{3a} + \frac{2}{a}\right\} \\
&< \frac{\sqrt{10}-2}{3} + \frac{4\sqrt{10}-2}{\sqrt{10}-2} \\
&< \frac{\sqrt{10}+1}{3} + 3\frac{\sqrt{10}}{\sqrt{10}-2}.
\end{aligned} \tag{4.179}$$

For $a < \frac{\sqrt{10}-2}{3}$ and using the highest possible marginal costs, (4.177) is obtained by

$$\bar{p} < a + 4 + \frac{2}{a}. \tag{4.180}$$

□

The bid function is shown for independently distributed marginal costs in Figure 4.16. Figure 4.17 presents bids for the strong affiliation with $a = 4$. Bids increase with marginal costs as well as with demand. The difference between the bids for independence and strong affiliation are shown in Figure 4.18. Negative values result in higher bids for independent distribution.

Ignoring the fact that the figures include the demand k, which is explicitly excluded in the analysis of this case, only in the case of independent distribution does the generator with the lowest marginal costs bid higher in the example. All other generators bid lower or the same for this distribution compared to the strong affiliation. In the example, if the distribution changes from independence to affiliation the greatest relative bid reduction is about 25%.

The figures also show that there is a jump in the bid function at the demand k. This demand represents the crossing from one competitive environment (case A) to the other (case B). It is interesting to see that a small change in demand leads to a different bidding strategy.

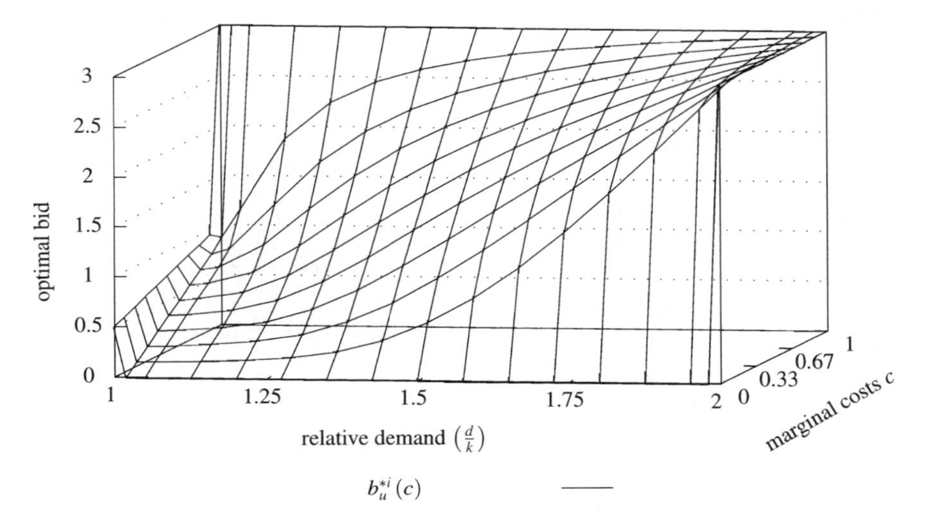

Fig. 4.16 Optimal bid function in a uniform-price auction for independently distributed costs (case B)

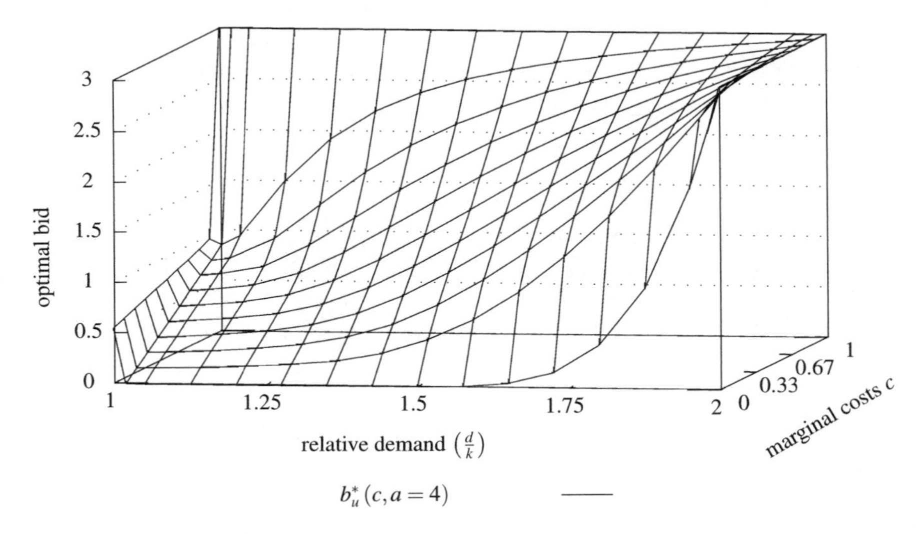

Fig. 4.17 Optimal bid function in a uniform-price auction for strong affiliation (case B)

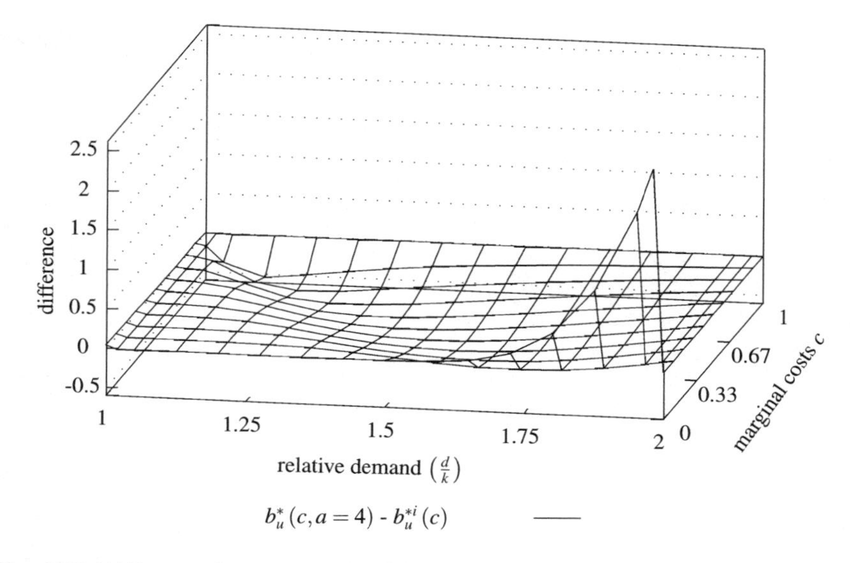

Fig. 4.18 Difference between optimal bids in a uniform-price auction (case B)

The following theorem relies on the general results of Theorem 4.1.

Theorem 4.16. *Suppose $d \in (k, 2k)$.*

(a) *The auction leads to production efficiency. Production costs are the lowest possible.*

(b) *The expected production costs per electricity unit are obtained by*

$$E[c_u] = 2\left[1-\frac{k}{d}\right]\bar{c} - 2\frac{k}{d}\int_{\underline{c}}^{\bar{c}} F(c)\,dc$$
$$+2\left[\frac{2k}{d}-1\right]\int_{\underline{c}}^{\bar{c}} F(c)\left\{c[f(c|c)+F_i(c|c)]+F(c|c)\right\}dc. \quad (4.181)$$

For independently distributed marginal costs, the expression reduces to

$$E\left[c_u^i\right] = \bar{c} + \left[\frac{2k}{d}-1\right]\int_{\underline{c}}^{\bar{c}} F(c)^2\,dc - \frac{2k}{d}\int_{\underline{c}}^{\bar{c}} F(c)\,dc. \quad (4.182)$$

Proof.

(a) Optimal bid function (4.143) increases with marginal costs and represents bids, which are accepted bids according to the auction rules. Hence, Theorem 4.1 applies, especially (a).

(b) Based on the same reason as given in (a), (4.6) and (4.7) apply. □

Theorem 4.17. *Suppose* $d \in (k, 2k)$.

(a) *The expected profit of a generator is obtained by*

$$\pi_u^*(c) = [d-k]F(c|c)\left[\bar{p} - \bar{c} + \int_c^{\bar{c}} \exp\left(\frac{2k-d}{d-k}\int_\tau^{\bar{c}} \frac{f(t|t)}{F(t|t)}\,dt\right) d\tau\right]$$
$$\times \exp\left(\frac{2k-d}{d-k}\int_c^{\bar{c}} \frac{(-)f(t|t)}{F(t|t)}\,dt\right) + k\int_c^{\bar{c}} [1 - F(\tau|c)]\,d\tau$$
$$+ k\int_c^{\bar{c}} f(\xi|c)\left[\bar{p} - \bar{c} + \int_\xi^{\bar{c}} \exp\left(\frac{2k-d}{d-k}\int_\tau^{\bar{c}} \frac{f(t|t)}{F(t|t)}\,dt\right) d\tau\right]$$
$$\times \exp\left(\frac{2k-d}{d-k}\int_\xi^{\bar{c}} \frac{(-)f(t|t)}{F(t|t)}\,dt\right) d\xi. \tag{4.183}$$

The expression for independently distributed marginal costs reduces to

$$\pi_u^{*i}(c) = [d-k][\bar{p} - c] + [2k-d]\int_c^{\bar{c}} [1 - F(\tau)]\,d\tau. \tag{4.184}$$

(b) *The expected contribution margin per capacity unit is determined by*

$$\pi_u^{u^*}(c) = \left[\frac{d}{k} - 1\right] F(c|c)\left[\bar{p} - \bar{c} + \int_c^{\bar{c}} \exp\left(\frac{2k-d}{d-k}\int_\tau^{\bar{c}} \frac{f(t|t)}{F(t|t)}\,dt\right) d\tau\right]$$
$$\times \exp\left(\frac{2k-d}{d-k}\int_c^{\bar{c}} \frac{(-)f(t|t)}{F(t|t)}\,dt\right) + \int_c^{\bar{c}} [1 - F(\tau|c)]\,d\tau$$
$$+ \int_c^{\bar{c}} f(\xi|c)\left[\bar{p} - \bar{c} + \int_\xi^{\bar{c}} \exp\left(\frac{2k-d}{d-k}\int_\tau^{\bar{c}} \frac{f(t|t)}{F(t|t)}\,dt\right) d\tau\right]$$
$$\times \exp\left(\frac{2k-d}{d-k}\int_\xi^{\bar{c}} \frac{(-)f(t|t)}{F(t|t)}\,dt\right) d\xi. \tag{4.185}$$

For independently distributed marginal costs, the expression reduces to

$$\pi_u^{u^{*i}}(c) = \left[\frac{d}{k} - 1\right][\bar{p} - c] + \left[2 - \frac{d}{k}\right] \int_c^{\bar{c}} [1 - F(\tau)]\, d\tau. \tag{4.186}$$

Proof.

(a) The expected profit, using (4.149), (4.2), and bid function (4.143), is given by

$$\begin{aligned}
\pi_u^*(c) =& [d-k]F(c|c)\,[b_u^*(c) - c] + k \int_c^{\bar{c}} f(\tau|c)\,[b_u^*(\tau) - \tau]\, d\tau \\
&+ k \int_c^{\bar{c}} f(\tau|c)\,[\tau - c]\, d\tau \\
=& [d-k]F(c|c) \left[\bar{p} - \bar{c} + \int_c^{\bar{c}} e^{\frac{2k-d}{d-k} \int_\tau^{\bar{c}} \frac{f(t|t)}{F(t|t)} dt} d\tau\right] e^{\frac{2k-d}{d-k} \int_c^{\bar{c}} \frac{(-)f(t|t)}{F(t|t)} dt} \\
&+ k \int_c^{\bar{c}} f(\xi|c) \left[\bar{p} - \bar{c} + \int_\xi^{\bar{c}} e^{\frac{2k-d}{d-k} \int_\tau^{\bar{c}} \frac{f(t|t)}{F(t|t)} dt} d\tau\right] e^{\frac{2k-d}{d-k} \int_\xi^{\bar{c}} \frac{(-)f(t|t)}{F(t|t)} dt} d\xi \\
&+ k\,F(\tau|c)\,[\tau - c]\Big|_c^{\bar{c}} - k \int_c^{\bar{c}} F(\tau|c)\, d\tau \\
=& [d-k]F(c|c) \left[\bar{p} - \bar{c} + \int_c^{\bar{c}} e^{\frac{2k-d}{d-k} \int_\tau^{\bar{c}} \frac{f(t|t)}{F(t|t)} dt} d\tau\right] e^{\frac{2k-d}{d-k} \int_c^{\bar{c}} \frac{(-)f(t|t)}{F(t|t)} dt} \\
&+ k \int_c^{\bar{c}} f(\xi|c) \left[\bar{p} - \bar{c} + \int_\xi^{\bar{c}} e^{\frac{2k-d}{d-k} \int_\tau^{\bar{c}} \frac{f(t|t)}{F(t|t)} dt} d\tau\right] e^{\frac{2k-d}{d-k} \int_\xi^{\bar{c}} \frac{(-)f(t|t)}{F(t|t)} dt} d\xi \\
&+ k \int_c^{\bar{c}} [1 - F(\tau|c)]\, d\tau.
\end{aligned}$$

The expression for independently distributed marginal costs, using (3.11), (3.12), and (3.3), can be rewritten to

$$\pi_u^{*i}(c) = [d-k]F(c)\left[\bar{p}-\bar{c}+\int_c^{\bar{c}} e^{\frac{2k-d}{d-k}\int_\tau^{\bar{c}}\frac{f(t)}{F(t)}dt}\,d\tau\right] e^{\frac{2k-d}{d-k}\int_c^{\bar{c}}\frac{(-)f(t)}{F(t)}dt}$$

$$+k\int_c^{\bar{c}} f(\xi)\left[\bar{p}-\bar{c}+\int_\xi^{\bar{c}} e^{\frac{2k-d}{d-k}\int_\tau^{\bar{c}}\frac{f(t)}{F(t)}dt}\,d\tau\right] e^{\frac{2k-d}{d-k}\int_\xi^{\bar{c}}\frac{(-)f(t)}{F(t)}dt}\,d\xi$$

$$+k\int_c^{\bar{c}}[1-F(\tau)]\,d\tau$$

$$= [d-k]F(c)^{\frac{k}{d-k}}\left[\bar{p}-\bar{c}+\int_c^{\bar{c}} F(\tau)^{(-)\frac{2k-d}{d-k}}\,d\tau\right] + k\int_c^{\bar{c}}[1-F(\tau)]\,d\tau$$

$$+k\int_c^{\bar{c}} f(\xi)F(\xi)^{\frac{2k-d}{d-k}}\left[\bar{p}-\bar{c}+\int_\xi^{\bar{c}} F(\tau)^{(-)\frac{2k-d}{d-k}}\,d\tau\right]d\xi.$$

For simplicity, the following term is used

$$\int f(\tau)F(\tau)^{\frac{2k-d}{d-k}}\,d\tau$$

$$= \frac{d-k}{2k-d}\int\left[\frac{2k-d}{d-k}f(\tau)F(\tau)^{\frac{3k-2d}{d-k}}\right]F(\tau)\,d\tau$$

$$= \frac{d-k}{2k-d}\left[F(\tau)^{\frac{k}{d-k}} - \int f(\tau)F(\tau)^{\frac{2k-d}{d-k}}\,d\tau\right]$$

$$= \left[\frac{d}{k}-1\right]F(\tau)^{\frac{k}{d-k}}$$

to derive

$$\int_c^{\bar{c}} f(\xi)F(\xi)^{\frac{2k-d}{d-k}}\left[\bar{p}-\bar{c}+\int_\xi^{\bar{c}} F(\tau)^{(-)\frac{2k-d}{d-k}}\,d\tau\right]d\xi$$

$$= \left[\frac{d}{k}-1\right]\left\{F(\xi)^{\frac{k}{d-k}}\left[\bar{p}-\bar{c}+\int_\xi^{\bar{c}} F(\tau)^{(-)\frac{2k-d}{d-k}}\,d\tau\right]\Bigg|_c^{\bar{c}} + \int_c^{\bar{c}} F(\xi)\,d\xi\right\}$$

$$= \left[\frac{d}{k}-1\right]\left\{\bar{p}-\bar{c}-F(c)^{\frac{k}{d-k}}\left[\bar{p}-\bar{c}+\int_c^{\bar{c}} F(\tau)^{(-)\frac{2k-d}{d-k}}\,d\tau\right] + \int_c^{\bar{c}} F(\tau)\,d\tau\right\}.$$

This is plugged into the expression of expected profit above to get

$$\begin{aligned}
\pi_u^{*i}(c) &= [d-k]F(c)^{\frac{k}{d-k}}\left[\bar{p}-\bar{c}+\int_c^{\bar{c}} F(\tau)^{(-)\frac{2k-d}{d-k}}\,d\tau\right]+[d-k][\bar{p}-\bar{c}] \\
&\quad -[d-k]\left\{F(c)^{\frac{k}{d-k}}\left[\bar{p}-\bar{c}+\int_c^{\bar{c}} F(\tau)^{(-)\frac{2k-d}{d-k}}\,d\tau\right]-\int_c^{\bar{c}} F(\tau)\,d\tau\right\} \\
&\quad +k\left[\bar{c}-c-\int_c^{\bar{c}} F(\tau)\,d\tau\right] \\
&= [d-k][\bar{p}-\bar{c}]+k[\bar{c}-c]-[2k-d]\int_c^{\bar{c}} F(\tau)\,d\tau \\
&= [d-k][\bar{p}-c]+[2k-d]\int_c^{\bar{c}}[1-F(\tau)]\,d\tau.
\end{aligned}$$

(b) The expected contribution margin for an electricity unit, using (4.183), is given by

$$\begin{aligned}
\pi_u^{u^*}(c) &= \frac{\pi_u^*(c)}{k} \\
&= \left[\frac{d}{k}-1\right]F(c|c)\left[\bar{p}-\bar{c}+\int_c^{\bar{c}} \mathrm{e}^{\frac{2k-d}{d-k}\int_\tau^{\bar{c}}\frac{f(t|t)}{F(t|t)}\,dt}\,d\tau\right]\mathrm{e}^{\frac{2k-d}{d-k}\int_c^{\bar{c}}\frac{(-)f(t|t)}{F(t|t)}\,dt} \\
&\quad +\int_c^{\bar{c}} f(\xi|c)\left[\bar{p}-\bar{c}+\int_\xi^{\bar{c}} \mathrm{e}^{\frac{2k-d}{d-k}\int_\tau^{\bar{c}}\frac{f(t|t)}{F(t|t)}\,dt}\,d\tau\right]\mathrm{e}^{\frac{2k-d}{d-k}\int_\xi^{\bar{c}}\frac{(-)f(t|t)}{F(t|t)}\,dt}\,d\xi \\
&\quad +\int_c^{\bar{c}}[1-F(\tau|c)]\,d\tau.
\end{aligned}$$

For independently distributed marginal costs, the expression, using (4.184), is obtained by

$$\begin{aligned}
\pi_u^{u^*}(c) &= \frac{\pi_u^{*i}(c)}{k} \\
&= \left[\frac{d}{k}-1\right][\bar{p}-c]+\left[2-\frac{d}{k}\right]\int_c^{\bar{c}}[1-F(\tau)]\,d\tau.
\end{aligned}$$

□

Corollary 4.8. *Suppose $d \in (k, 2k)$.*

(a) *The expected profit and expected contribution margin per capacity unit are positive.*

(b) *The expected profit and expected contribution margin per capacity unit increase strictly with demand.*

(c) *The expected profit and expected contribution margin per capacity unit decrease strictly with marginal costs if the following condition for all marginal costs holds true*

$$\int_{c}^{\bar{c}} f_i(\xi|c)\,[b_u^*(\xi) - c]\,d\xi$$

$$< 1 + \left[2 - \frac{d}{k}\right] F(c|c) - \left[\frac{d}{k} - 1\right] F_i(c|c)\,[b_u^*(c) - c]. \tag{4.187}$$

This condition is always satisfied for independent distribution.

Proof.

(a) The expression of the expected profit (4.183), using bid function (4.143), can also be rewritten to

$$\pi_u^*(c) = [d-k]F(c|c)\,[b_u^*(c) - c] + k\int_{c}^{\bar{c}} f(\xi|c)\,[b_u^*(\xi) - \xi]\,d\xi$$

$$+k\int_{c}^{\bar{c}} [1 - F(\tau|c)]\,d\tau. \tag{4.188}$$

The first summand is for all except $\underline{c}$ positive because $d > k$, $F(c|c) > 0$ due to (3.5), and $b_u^*(c) - c > 0$ due to Corollary 4.7 (a). The term is zero for $\underline{c}$ because $F(\underline{c}|\underline{c}) = 0$. The remaining summands are zero for $\bar{c}$ and positive for the other marginal costs. Hence, the whole term is positive for all marginal costs.

The same arguments are used to show that the expected contribution margin is also positive because the factor $\frac{1}{k}$ is positive.

(b) The first derivative of the expected profit with respect to demand, using (4.188), is given by

$$\frac{\partial \pi_u^*(c)}{\partial d} = F(c|c)\,[b_u^*(c) - c] + [d-k]F(c|c)\,\frac{\partial b_u^*(c)}{\partial d}$$

$$+k\int_{c}^{\bar{c}} f(\tau|c)\,\frac{\partial b_u^*(\tau)}{\partial d}\,d\tau. \tag{4.189}$$

The first two summands are zero for $\underline{c}$ because $F(\underline{c}|\underline{c}) = 0$, see (3.5). The first summand is positive for all other marginal costs because $F(c|c) > 0$ and $b_u^*(c) - c > 0$ due to Corollary 4.7 (a).

The second summand is zero for $\bar{c}$ and positive for $c \in (\underline{c}, \bar{c})$ because of Corollary 4.7 (c). The last summand is obviously zero for $\bar{c}$. It is positive for $c \in [\underline{c}, \bar{c})$ because $f(\tau|c) > 0$, see (3.4), and due to Corollary 4.7 (c). Hence, the sum is positive for all marginal costs, i.e. the expected profit increases strictly with demand.

The slope for the expected contribution margin, using (4.185), is obtained by

$$\frac{\partial \pi_u^{u^*}(c)}{\partial d} = \frac{1}{k}\frac{\partial \pi_u^*(c)}{\partial d}. \tag{4.190}$$

The derivative is also positive due to the same arguments given for the expected profit.

(c) The first derivative of the expected profit, using (4.183), (4.143), (4.155), (4.1)–(4.3), and (3.5), is given by

$$\begin{aligned}
\frac{\partial \pi_u^*(c)}{\partial c} &= [d-k][f(c|c) + F_i(c|c)][b_u^*(c) - c] \\
&\quad + [d-k]F(c|c)[b_u^*(c)' - 1] + k\int_c^{\bar{c}} f_i(\xi|c)[b_u^*(\xi) - c]\,d\xi \\
&\quad - k\int_c^{\bar{c}} f(\xi|c)\,d\xi - kf(c|c)[b_u^*(c) - c] \\
&= [d-2k]f(c|c)[b_u^*(c) - c] + [d-k]F_i(c|c)[b_u^*(c) - c] \\
&\quad + [2k-d]f(c|c)[b_u^*(c) - c] - [d-k]F(c|c) \\
&\quad + k\int_c^{\bar{c}} f_i(\xi|c)\,b_u^*(\xi)\,d\xi - kc\int_c^{\bar{c}} f_i(\xi|c)\,d\xi - k\int_c^{\bar{c}} f(\xi|c)\,d\xi \\
&= [d-k]F_i(c|c)[b_u^*(c) - c] - [d-k]F(c|c) - k[1 - F(c|c)] \\
&\quad + k\int_c^{\bar{c}} f_i(\xi|c)[b_u^*(\xi) - c]\,d\xi \\
&= [d-k]F_i(c|c)[b_u^*(c) - c] - k + [2k-d]F(c|c) \\
&\quad + k\int_c^{\bar{c}} f_i(\xi|c)[b_u^*(\xi) - c]\,d\xi
\end{aligned}$$

The profit strictly decreases with marginal costs if the following holds true for all marginal costs

$$\frac{\partial \pi_u^*(c)}{\partial c} < 0$$

$$\Longleftrightarrow$$

$$\int_c^{\bar{c}} f_i(\xi | c)\,[b_u^*(\xi) - c]\,d\xi$$

$$< 1 + \left[2 - \frac{d}{k}\right] F(c|c) - \left[\frac{d}{k} - 1\right] F_i(c|c)\,[b_u^*(c) - c]. \tag{4.191}$$

The condition for independently distributed marginal costs, using (3.13), (3.14) and taking into account that $F(c|c) \leq 1$ due to (3.12), is given by

$$0 < 1 - \left[2 - \frac{d}{k}\right] F(c)$$

$$\Longleftrightarrow$$

$$1 > \left[2 - \frac{d}{k}\right] F(c)$$

$$\Longleftrightarrow$$

$$d > k.$$

This is valid given the demand range of this case. Hence, the expected profit decreases strictly with marginal costs for the independent distribution.

Similar arguments are found for the expected contribution margin because the factor $\frac{1}{k}$ is independent of c. □

Generators, even those with the highest marginal costs, have a strong incentive to stay in the market because expected profits are always positive according to Corollary 4.8 (a). The reason is that the loser can always sell some electricity and make money. That the expected profit increases with demand has two reasons. First, higher demand leads to the assured sale of a greater quantity of electricity in the case of losing the auction. Hence, the expected quantity sold increases with demand. The second reason is that bids increase with demand, see Corollary 4.7 (c). This is a consequence of the first reason.

Example 4.11. *Suppose the probability distribution given by Definition 3.5.*
(a) *The expected profit for independently distributed marginal costs is obtained by*

$$\pi_u^{*i}(c) = [d-k][\bar{p} - c] + [2k - d]\frac{[1-c]^2}{2}. \tag{4.192}$$

Given strong affiliation, the profit is determined by

$$\pi_u^*(c) = [d-k]\frac{2+ac^2}{2+ac}c\left\{\bar{p}-1+\int_c^1 \exp\left(\frac{2k-d}{d-k}\int_\tau^1 \frac{2[1+at^2]}{[2+at^2]t}dt\right)d\tau\right\}$$

$$\times \exp\left(\frac{2k-d}{d-k}\int_c^1 \frac{(-2)[1+at^2]}{[2+at^2]t}dt\right)$$

$$+2k\int_c^1 \frac{1+ac\xi}{2+ac}\left\{\bar{p}-1+\int_\xi^1 \exp\left(\frac{2k-d}{d-k}\int_\tau^1 \frac{2[1+at^2]}{[2+at^2]t}dt\right)d\tau\right\}$$

$$\times \exp\left(\frac{2k-d}{d-k}\int_\xi^1 \frac{(-2)[1+at^2]}{[2+at^2]t}dt\right)d\xi$$

$$+k\frac{[1-c]^2}{3}\left[2+c-\frac{1+2c}{2+ac}\right]. \tag{4.193}$$

(b) *The expected contribution margin per capacity unit for independent distribution is given by*

$$\pi_u^{u*i}(c) = \left[\frac{d}{k}-1\right][\bar{p}-c]+\left[2-\frac{d}{k}\right]\frac{[1-c]^2}{2}. \tag{4.194}$$

The expression for strong affiliation is obtained by

$$\pi_u^{u*}(c) = \left[\frac{d}{k}-1\right]c\frac{2+ac^2}{2+ac}\exp\left(\frac{2k-d}{d-k}\int_c^1 \frac{(-2)[1+at^2]}{[2+at^2]t}dt\right)$$

$$\times\left\{\bar{p}-1+\int_c^1 \exp\left(\frac{2k-d}{d-k}\int_\tau^1 \frac{2[1+at^2]}{[2+at^2]t}dt\right)d\tau\right\}$$

$$+2\int_c^1 \frac{1+ac\xi}{2+ac}\left\{\bar{p}-1+\int_\xi^1 \exp\left(\frac{2k-d}{d-k}\int_\tau^1 \frac{2[1+at^2]}{[2+at^2]t}dt\right)d\tau\right\}$$

$$\times \exp\left(\frac{2k-d}{d-k}\int_\xi^1 \frac{(-2)[1+at^2]}{[2+at^2]t}dt\right)d\xi$$

$$+\frac{[1-c]^2}{3}\left[2+c-\frac{1+2c}{2+ac}\right]. \tag{4.195}$$

Proof.

(a) The derivation of the profit uses following term

$$\begin{aligned}
\int_c^1 F(\tau|c)\,d\tau &= \int_c^1 \frac{1+\frac{a}{2}c\tau}{1+\frac{a}{2}c}\,\tau\,d\tau \\
&= \frac{1}{2+ac}\left[\tau^2+\frac{a}{3}c\tau^3\right]\Bigg|_c^1 \\
&= \frac{1-c}{2+ac}\left\{1+c+\frac{a}{3}[1+c+c^2]c\right\} \\
&= \begin{cases} \dfrac{1-c^2}{2} & : a=0 \\ \dfrac{1-c}{2+ac}\left\{1+c+\dfrac{a}{3}[1+c+c^2]c\right\} & : a>0. \end{cases}
\end{aligned} \tag{4.196}$$

The expected profit for independently distributed marginal costs, using (4.184) and taking into account (3.12) in combination with (4.196), is obtained by

$$\begin{aligned}
\pi_u^{*i}(c) &= [d-k][\bar{p}-c]+[2k-d]\left[\int_c^1 d\tau-\frac{1-c^2}{2}\right] \\
&= [d-k][\bar{p}-c]+[2k-d]\left[1-c-\frac{1-c^2}{2}\right] \\
&= [d-k][\bar{p}-c]+[2k-d]\frac{[1-c]^2}{2}.
\end{aligned}$$

The expected profit for strong affiliation, using (4.183), (3.26), (3.27), and (4.196), is obtained by

$$\pi_u^*(c) = [d-k]\frac{1+\frac{a}{2}c^2}{1+\frac{a}{2}c}c\left[\bar{p}-1+\int_c^1 e^{\frac{2k-d}{d-k}\int_\tau^1 \frac{\frac{1+at^2}{1+\frac{a}{2}t}}{\frac{1+\frac{a}{2}t^2}{1+\frac{a}{2}t}t}dt} d\tau\right]$$

$$\times e^{\frac{2k-d}{d-k}\int_c^1 \frac{(-)\frac{1+at^2}{1+\frac{a}{2}t}}{\frac{1+\frac{a}{2}t^2}{1+\frac{a}{2}t}t}dt} + k\int_c^1 \frac{1+ac\xi}{1+\frac{a}{2}c} e^{\frac{2k-d}{d-k}\int_\xi^1 \frac{(-)\frac{1+at^2}{1+\frac{a}{2}t}}{\frac{1+\frac{a}{2}t^2}{1+\frac{a}{2}t}t}dt}$$

$$\times\left[\bar{p}-1+\int_\xi^1 e^{\frac{2k-d}{d-k}\int_\tau^1 \frac{\frac{1+at^2}{1+\frac{a}{2}t}}{\frac{1+\frac{a}{2}t^2}{1+\frac{a}{2}t}t}dt} d\tau\right]d\xi$$

$$+k[1-c]-k\frac{1-c}{2+ac}\left\{1+c+\frac{a}{3}[1+c+c^2]c\right\}$$

$$= k\frac{[1-c]^2}{3}\left[2+c-\frac{1+2c}{2+ac}\right]+[d-k]\frac{2+ac^2}{2+ac}c$$

$$\times\left\{\bar{p}-1+\int_c^1 e^{\frac{2k-d}{d-k}\int_\tau^1 \frac{2[1+at^2]}{[2+at^2]t}dt} d\tau\right\} e^{\frac{2k-d}{d-k}\int_c^1 \frac{(-2)[1+at^2]}{[2+at^2]t}dt}$$

$$+2k\int_c^1 \frac{1+ac\xi}{2+ac}\left\{\bar{p}-1+\int_\xi^1 e^{\frac{2k-d}{d-k}\int_\tau^1 \frac{2[1+at^2]}{[2+at^2]t}dt} d\tau\right\}$$

$$\times e^{\frac{2k-d}{d-k}\int_\xi^1 \frac{(-2)[1+at^2]}{[2+at^2]t}dt} d\xi.$$

(b) The expected contribution margin for strong affiliation is derived from the definition of (4.185) using (4.193)

$$\pi_u^{u^*}(c) = \frac{\pi_u^*(c)}{k}$$

$$= \left[\frac{d}{k} - 1\right] \frac{2+ac^2}{2+ac} c \left\{ \bar{p} - 1 + \int_c^1 e^{\frac{2k-d}{d-k} \int_\tau^1 \frac{2[1+at^2]}{[2+at^2]t} dt} d\tau \right\}$$

$$\times e^{\frac{2k-d}{d-k} \int_c^1 \frac{(-2)[1+at^2]}{[2+at^2]t} dt} + 2 \int_c^1 \frac{1+ac\xi}{2+ac} e^{\frac{2k-d}{d-k} \int_\xi^1 \frac{(-2)[1+at^2]}{[2+at^2]t} dt}$$

$$\times \left\{ \bar{p} - 1 + \int_\xi^1 e^{\frac{2k-d}{d-k} \int_\tau^1 \frac{2[1+at^2]}{[2+at^2]t} dt} d\tau \right\} d\xi$$

$$+ \frac{[1-c]^2}{3} \left[2 + c - \frac{1+2c}{2+ac}\right].$$

The expression for independent distribution, using (4.192), is obtained by

$$\pi_u^{u*i}(c) = \frac{\pi_u^{*i}(c)}{k}$$

$$= \left[\frac{d}{k} - 1\right] [\bar{p} - c] + \left[2 - \frac{d}{k}\right] \frac{[1-c]^2}{2}.$$

□

The expected contribution margin for independently distributed marginal costs is shown in Figure 4.19. The graph for strong affiliation is presented in Figure 4.20 with $a = 4$. The expected contribution margin of the example increases with demand but decreases with marginal costs.

The difference between the expected contribution margin for strong affiliation and independently distributed marginal costs is shown in Figure 4.21. Negative values signify that the value for the independent distribution is greater.

For every demand in this case there are generators who receive higher expected contribution margins with independent marginal costs, while at the same time other generators earn higher profits given strong affiliation. The relative excess profit between both distribution settings is shown in Figure 4.22. The greatest relative profit decline for the change to strong affiliation is about 16% in the example.

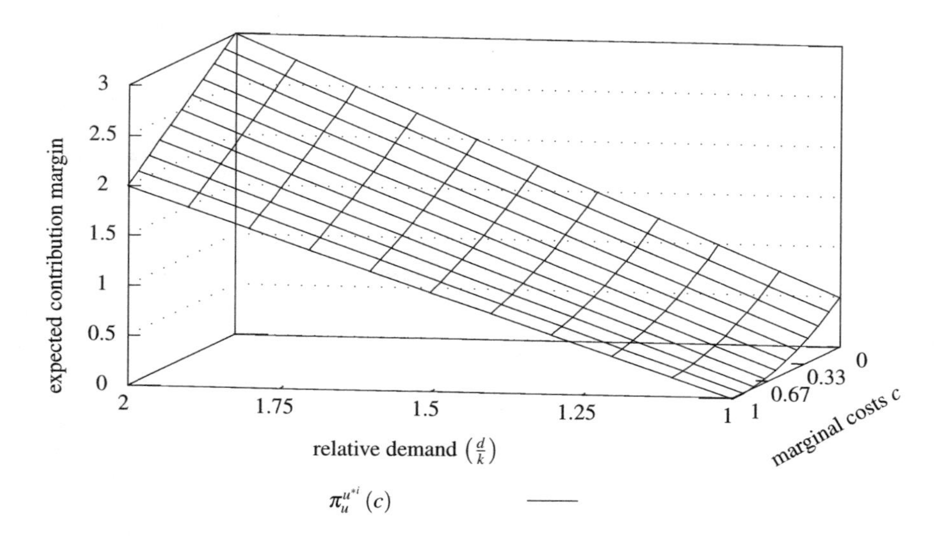

Fig. 4.19 Expected contribution margin in a uniform-price auction for independently distributed costs (case B)

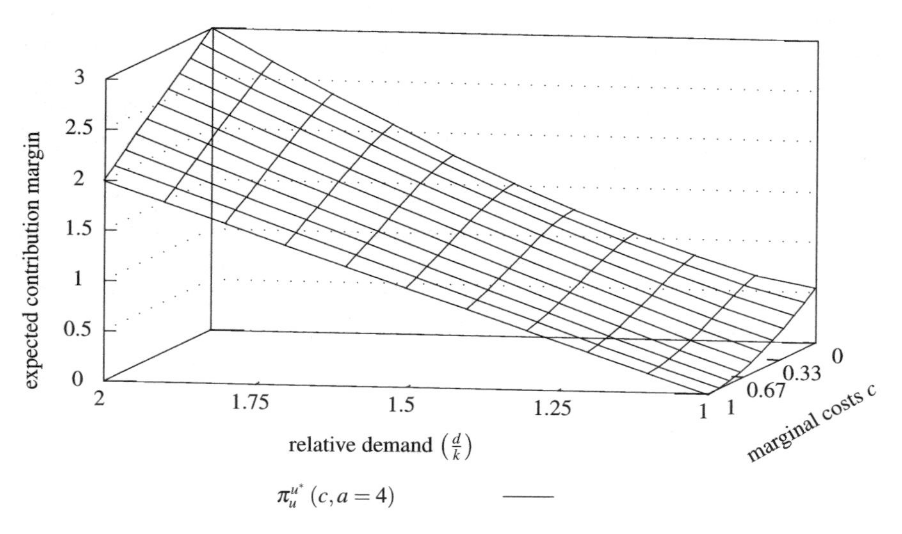

Fig. 4.20 Expected contribution margin in a uniform-price auction for strong affiliation (case B)

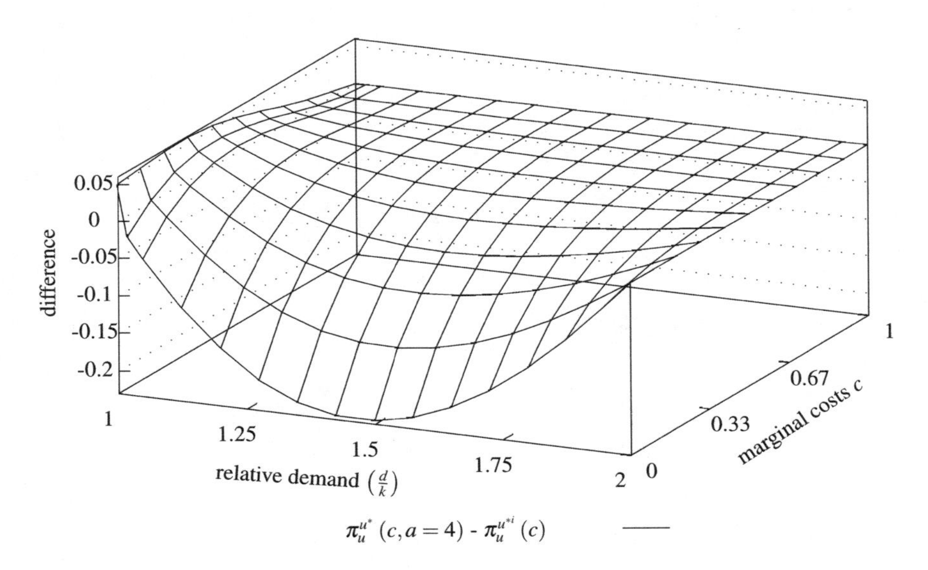

Fig. 4.21 Difference in contribution margins in a uniform-price auction (case B)

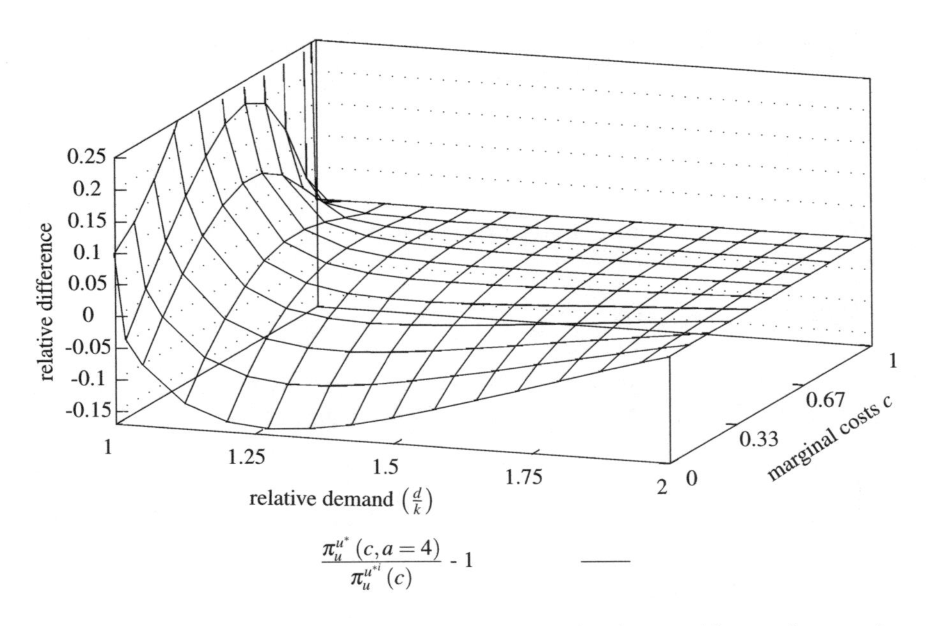

Fig. 4.22 Relative difference in contribution margins in a uniform-price auction (case B)

Theorem 4.18. *Suppose $d \in (k, 2k)$. The expected uniform price is given by*

$$E[p_u] = 2\bar{p} - 2\int_{\underline{c}}^{\bar{c}} F(c)\left[\frac{kf(c|c)}{d-k} + F_i(c|c)\right]\exp\left((-)\frac{2k-d}{d-k}\int_{c}^{\bar{c}}\frac{f(t|t)}{F(t|t)}dt\right)$$
$$\times\left[\bar{p} - \bar{c} + \int_{c}^{\bar{c}}\exp\left(\frac{2k-d}{d-k}\int_{\tau}^{\bar{c}}\frac{f(t|t)}{F(t|t)}dt\right)d\tau\right]dc$$
$$-2\int_{\underline{c}}^{\bar{c}} cF(c)\left[f(c|c) + F_i(c|c)\right]dc. \tag{4.197}$$

For independently distributed marginal costs, the expression reduces to

$$E\left[p_u^i\right] = 2\left[1 - \frac{k}{d}\right]\bar{p} + \left[\frac{2k}{d} - 1\right]\left[\bar{c} - \int_{\underline{c}}^{\bar{c}} F(c)^2\,dc\right]. \tag{4.198}$$

Proof.

The bid of the losing generator determines the uniform price. With $j \neq i$, the expected uniform price is therefore given by

$$E[p_u] = E\left[b_u^*(c_i)\,1_{\{b_u^*(c_i) > b_u^*(c_j)\}} + b_u^*(c_j)\,1_{\{b_u^*(c_j) > b_u^*(c_i)\}}\right].$$

Due to identical bid functions, the expected uniform price is obtained by

$$E[p_u] = 2E\left[b_u^*(c_i)\,1_{\{b_u^*(c_i) > b_u^*(c_j)\}}\right].$$

This can be simplified because the bid function is strictly increasing in c. Using (3.5), (4.155), (4.2), and (4.3), the function is then obtained by

$$E[p_u] = 2E\left[b_u^*(c_i)\,1_{\{c_i > c_j\}}\right]$$

$$= 2\int_{\underline{c}}^{\bar{c}}\int_{\underline{c}}^{c} b_u^*(c)\,f(\tau, c)\,d\tau\,dc$$

$$= 2\int_{\underline{c}}^{\bar{c}} b_u^*(c)\, f(c)\, F(c|c)\, dc \tag{4.199}$$

$$= 2\,[b_u^*(c) - c]F(c)\,F(c|c)\Big|_{\underline{c}}^{\bar{c}} - 2\int_{\underline{c}}^{\bar{c}} [b_u^*(c)' - 1]F(c)\,F(c|c)\,dc$$

$$-2\int_{\underline{c}}^{\bar{c}} [b_u^*(c) - c]F(c)\,[f(c|c) + F_i(c|c)]\,dc + 2\,cF(c)\,F(c|c)\Big|_{\underline{c}}^{\bar{c}}$$

$$-2\int_{\underline{c}}^{\bar{c}} F(c)\,F(c|c)\,dc - 2\int_{\underline{c}}^{\bar{c}} cF(c)\,[f(c|c) + F_i(c|c)]\,dc$$

$$= 2\bar{p} - 2\int_{\underline{c}}^{\bar{c}} cF(c)\,[f(c|c) + F_i(c|c)]\,dc$$

$$-2\int_{\underline{c}}^{\bar{c}} [b_u^*(c) - c]F(c)\left[\frac{k}{d-k} f(c|c) + F_i(c|c)\right] dc.$$

Substituting (4.143) for the bid function, the expected uniform price is given by

$$E[p_u] = 2\bar{p} - 2\int_{\underline{c}}^{\bar{c}} cF(c)\,[f(c|c) + F_i(c|c)]\,dc$$

$$-2\int_{\underline{c}}^{\bar{c}} F(c)\left[\frac{kf(c|c)}{d-k} + F_i(c|c)\right]\left[\bar{p} - \bar{c} + \int_{c}^{\bar{c}} \mathrm{e}^{\frac{2k-d}{d-k}\int_{\tau}^{\bar{c}} \frac{f(t|t)}{F(t|t)}\,dt}\, d\tau\right]$$

$$\times \mathrm{e}^{(-)\frac{2k-d}{d-k}\int_{c}^{\bar{c}} \frac{f(t|t)}{F(t|t)}\,dt}\, dc.$$

Using (3.11), (3.12), (3.14), and (4.11), the expression can be simplified for independently distributed marginal costs to

$$E\left[p_u^i\right] = 2\bar{p} - 2\int_{\underline{c}}^{\bar{c}} cf(c)\,F(c)\,dc - \frac{2k}{d-k}\int_{\underline{c}}^{\bar{c}} f(c)\,F(c)$$

$$\times \left[\bar{p} - \bar{c} + \int_{c}^{\bar{c}} \mathrm{e}^{\frac{2k-d}{d-k}\int_{\tau}^{\bar{c}} \frac{f(t)}{F(t)}\,dt}\, d\tau\right] \mathrm{e}^{(-)\frac{2k-d}{d-k}\int_{c}^{\bar{c}} \frac{f(t)}{F(t)}\,dt}\, dc$$

$$= 2\bar{p} - \bar{c} + \int_{\underline{c}}^{\bar{c}} F(c)^2\, dc - \frac{2k}{d-k}[\bar{p} - \bar{c}] \int_{\underline{c}}^{\bar{c}} f(c) F(c) \mathrm{e}^{(-)\frac{2k-d}{d-k} \int_{c}^{\bar{c}} \frac{f(t)}{F(t)}\, dt}\, dc$$

$$- \frac{2k}{d-k} \int_{\underline{c}}^{\bar{c}} f(c) F(c) \int_{c}^{\bar{c}} \mathrm{e}^{(-)\frac{2k-d}{d-k} \int_{c}^{\tau} \frac{f(t)}{F(t)}\, dt}\, d\tau\, dc.$$

The following two expressions are derived to simplify the proof. Due to

$$\frac{\partial}{\partial c} \mathrm{e}^{(-)\frac{2k-d}{d-k} \int_{c}^{\bar{c}} \frac{f(t)}{F(t)}\, dt} = \frac{2k-d}{d-k} \frac{f(c)}{F(c)} \mathrm{e}^{(-)\frac{2k-d}{d-k} \int_{c}^{\bar{c}} \frac{f(t)}{F(t)}\, dt},$$

the following term, using (3.3), is determined by

$$\int_{\underline{c}}^{\bar{c}} f(c) F(c) \mathrm{e}^{(-)\frac{2k-d}{d-k} \int_{c}^{\bar{c}} \frac{f(t)}{F(t)}\, dt}\, dc$$

$$= \frac{d-k}{2k-d} \int_{\underline{c}}^{\bar{c}} \frac{2k-d}{d-k} \frac{f(c)}{F(c)} \mathrm{e}^{(-)\frac{2k-d}{d-k} \int_{c}^{\bar{c}} \frac{f(t)}{F(t)}\, dt} F(c)^2\, dc$$

$$= \frac{d-k}{2k-d} \left[\mathrm{e}^{(-)\frac{2k-d}{d-k} \int_{c}^{\bar{c}} \frac{f(t)}{F(t)}\, dt} F(c)^2 \Big|_{\underline{c}}^{\bar{c}} - 2 \int_{\underline{c}}^{\bar{c}} f(c) F(c) \mathrm{e}^{(-)\frac{2k-d}{d-k} \int_{c}^{\bar{c}} \frac{f(t)}{F(t)}\, dt}\, dc \right]$$

$$= \frac{d-k}{2k-d} \left[1 - 2 \int_{\underline{c}}^{\bar{c}} f(c) F(c) \mathrm{e}^{(-)\frac{2k-d}{d-k} \int_{c}^{\bar{c}} \frac{f(t)}{F(t)}\, dt}\, dc \right]$$

$$= 1 - \frac{k}{d}. \tag{4.200}$$

Based on

$$\frac{\partial}{\partial c} \int_{c}^{\bar{c}} \mathrm{e}^{(-)\frac{2k-d}{d-k} \int_{c}^{\tau} \frac{f(t)}{F(t)}\, dt}\, d\tau$$

$$= \frac{2k-d}{d-k} \frac{f(c)}{F(c)} \int_{c}^{\bar{c}} \mathrm{e}^{(-)\frac{2k-d}{d-k} \int_{c}^{\tau} \frac{f(t)}{F(t)}\, dt}\, d\tau - \mathrm{e}^{(-)\frac{2k-d}{d-k} \int_{c}^{c} \frac{f(t)}{F(t)}\, dt}$$

$$= \frac{2k-d}{d-k}\frac{f(c)}{F(c)}\int_{c}^{\bar{c}} \mathrm{e}^{(-)\frac{2k-d}{d-k}\int_{c}^{\tau}\frac{f(t)}{F(t)}dt}\,d\tau - 1,$$

the following expression is obtained by

$$\int_{\underline{c}}^{\bar{c}} f(c)F(c)\int_{c}^{\bar{c}} \mathrm{e}^{(-)\frac{2k-d}{d-k}\int_{c}^{\tau}\frac{f(t)}{F(t)}dt}\,d\tau dc$$

$$= \frac{d-k}{2k-d}\int_{\underline{c}}^{\bar{c}} F(c)^2\left[\frac{2k-d}{d-k}\frac{f(c)}{F(c)}\int_{c}^{\bar{c}} \mathrm{e}^{(-)\frac{2k-d}{d-k}\int_{c}^{\tau}\frac{f(t)}{F(t)}dt}\,d\tau - 1\right]dc$$

$$+\frac{d-k}{2k-d}\int_{\underline{c}}^{\bar{c}} F(c)^2\,dc$$

$$= \frac{d-k}{2k-d}\int_{\underline{c}}^{\bar{c}} F(c)^2\,dc + \frac{d-k}{2k-d}F(c)^2\left[\int_{c}^{\bar{c}} \mathrm{e}^{(-)\frac{2k-d}{d-k}\int_{c}^{\tau}\frac{f(t)}{F(t)}dt}\,d\tau\right]\Bigg|_{\underline{c}}^{\bar{c}}$$

$$-2\frac{d-k}{2k-d}\int_{\underline{c}}^{\bar{c}} f(c)F(c)\int_{c}^{\bar{c}} \mathrm{e}^{(-)\frac{2k-d}{d-k}\int_{c}^{\tau}\frac{f(t)}{F(t)}dt}\,d\tau dc$$

$$= \left[1-\frac{k}{d}\right]\left[F(\bar{c})^2\int_{\bar{c}}^{\bar{c}} \mathrm{e}^{\frac{2k-d}{d-k}\int_{\tau}^{\bar{c}}\frac{f(t)}{F(t)}dt}\,d\tau - F(\underline{c})^2\int_{\underline{c}}^{\bar{c}} \mathrm{e}^{(-)\frac{2k-d}{d-k}\int_{\underline{c}}^{\tau}\frac{f(t)}{F(t)}dt}\,d\tau\right]$$

$$+\left[1-\frac{k}{d}\right]\int_{\underline{c}}^{\bar{c}} F(c)^2\,dc$$

$$= \left[1-\frac{k}{d}\right]\int_{\underline{c}}^{\bar{c}} F(c)^2\,dc. \tag{4.201}$$

Equations (4.200) and (4.201) are used to get

$$E\left[p_u^i\right] = 2\bar{p} - \bar{c} + \int_{\underline{c}}^{\bar{c}} F(c)^2\,dc - \frac{2k}{d}[\bar{p}-\bar{c}] - \frac{2k}{d}\int_{\underline{c}}^{\bar{c}} F(c)^2\,dc$$

$$= 2\left[1-\frac{k}{d}\right]\bar{p} + \left[\frac{2k}{d}-1\right]\left[\bar{c} - \int_{\underline{c}}^{\bar{c}} F(c)^2\,dc\right].$$

□

Corollary 4.9. *Suppose $d \in (k, 2k)$.*

(a) *The expected uniform price is greater than the expected production costs.*
(b) *The expected uniform price increases strictly with demand.*

Proof.

(a) Suppose $c_i \geq c_j \, \forall \, c_i, c_j \in [\underline{c}, \bar{c}]$. The uniform price is the highest bid according to the auction rules. Hence, it is $b_u^*(c_i)$. The expected production costs (4.6) are the expected average of the realized production costs. If the difference in the highest bid for all combinations of marginal costs is greater than the average production costs then the expected uniform price is greater than the expected production costs. The following condition must therefore hold true, taking into account that the generator with the lowest marginal costs c_j produces k electricity units while the other sells $[d-k]$

$$b_u^*(c_i) > \frac{k}{d} c_j + \frac{d-k}{d} c_i.$$

This is true if the following is valid because $c_j \leq c_i$ as presumed

$$\begin{aligned} b_u^*(c_i) &> \frac{k}{d} c_i + \frac{d-k}{d} c_i \\ &> c_i. \end{aligned}$$

This relation was proved by Corollary 4.7 (a). Therefore, the expected uniform price is greater than the expected production costs.

(b) The first derivative of the uniform price, using (4.199) instead of (4.197), is obtained by

$$\frac{\partial E[p_u]}{\partial d} = 2 \int_{\underline{c}}^{\bar{c}} \frac{\partial b_u^*(c)}{\partial d} f(c) F(c|c) \, dc. \tag{4.202}$$

The derivative is positive if for all marginal costs but $\bar{c}$ the following holds true because the probability density function is positive due to Assumption 3.7

$$\frac{\partial b_u^*(c)}{\partial d} > 0.$$

This was proved by Corollary 4.7 (c). Hence, the first derivative is positive and the expected uniform price increases strictly with demand. □

The expected uniform price is higher than the expected production costs for one electricity unit, see Corollary 4.9 (a). This is the consequence of Assumption 3.2, which presumes that a generator should not suffer a loss. That the expected uniform price increases with demand according to Corollary 4.9 (b) is not surprising because bids increase with demand as Corollary 4.7 (c) showed.

Example 4.12.

Suppose the probability distribution given by Definition 3.5. The expected uniform price for independently distributed marginal costs is given by

$$E\left[p_u^i\right] = 2\left[1-\frac{k}{d}\right]\bar{p}+\frac{2}{3}\left[\frac{2k}{d}-1\right]. \tag{4.203}$$

For strong affiliation, the uniform price is determined by

$$\begin{aligned} E\left[p_u\right] = {} & 2\bar{p}-\frac{4}{5}+\frac{2}{4+a}\left[\frac{1}{10}+\frac{2}{3a}-\frac{4}{a^2}-\frac{16}{a^3}+16\frac{2+a}{a^4}\ln\left(1+\frac{a}{2}\right)\right] \\ & -\frac{4}{4+a}\int_0^1 \frac{4+ac}{2+ac}\left[k\frac{1+ac^2}{d-k}-ac\frac{1-c}{2+ac}\right]c \\ & \times\left\{\bar{p}-1+\int_c^1 \exp\left(\frac{2k-d}{d-k}\int_\tau^1 \frac{2[1+at^2]}{[2+at^2]t}\,dt\right)d\tau\right\} \\ & \times \exp\left(\frac{2k-d}{d-k}\int_c^1 \frac{(-2)[1+at^2]}{[2+at^2]t}\,dt\right)dc. \end{aligned} \tag{4.204}$$

Proof.

According to (4.198), the expected uniform price for independent distribution, using (3.25), is given by

$$\begin{aligned} E\left[p_u^i\right] &= 2\left[1-\frac{k}{d}\right]\bar{p}+\left[\frac{2k}{d}-1\right]\left[1-\int_0^1 c^2\,dc\right] \\ &= 2\left[1-\frac{k}{d}\right]\bar{p}+\frac{2}{3}\left[\frac{2k}{d}-1\right]. \end{aligned}$$

Using (4.197), (3.25), (3.26), (3.27), (3.32), and (4.17), the uniform price for strong affiliation is obtained by

$$E[p_u] = 2\bar{p} - \frac{4}{5} + \frac{2}{4+a}\left[\frac{1}{10} + \frac{2}{3a} - \frac{4}{a^2} - \frac{16}{a^3} + 16\frac{2+a}{a^4}\ln\left(1+\frac{a}{2}\right)\right]$$

$$-2[\bar{p}-1]\int_0^1 \frac{1+\frac{a}{4}c}{1+\frac{a}{4}} c \left\{\frac{k\frac{1+ac^2}{1+\frac{a}{2}c}}{d-k} - \frac{a[1-c]c}{2\left[1+\frac{a}{2}c\right]^2}\right\} \mathrm{e}^{(-)\frac{2k-d}{d-k}\int_c^1 \frac{\frac{1+at^2}{1+\frac{a}{2}t}}{\frac{1+\frac{a}{2}t^2}{1+\frac{a}{2}t}t} dt} dc$$

$$-2\int_0^1 \frac{1+\frac{a}{4}c}{1+\frac{a}{4}} c \left\{\frac{k\frac{1+ac^2}{1+\frac{a}{2}c}}{d-k} - \frac{a[1-c]c}{2\left[1+\frac{a}{2}c\right]^2}\right\} \int_c^1 \mathrm{e}^{(-)\frac{2k-d}{d-k}\int_c^\tau \frac{\frac{1+at^2}{1+\frac{a}{2}t}}{\frac{1+\frac{a}{2}t^2}{1+\frac{a}{2}t}t} dt} d\tau dc$$

$$= 2\bar{p} - \frac{4}{5} + \frac{2}{4+a}\left[\frac{1}{10} + \frac{2}{3a} - \frac{4}{a^2} - \frac{16}{a^3} + 16\frac{2+a}{a^4}\ln\left(1+\frac{a}{2}\right)\right]$$

$$-\frac{4}{4+a}\int_0^1 \frac{4+ac}{2+ac}\left[k\frac{1+ac^2}{d-k} - ac\frac{1-c}{2+ac}\right] c$$

$$\times \left\{\bar{p} - 1 + \int_c^1 \mathrm{e}^{\frac{2k-d}{d-k}\int_\tau^1 \frac{2[1+at^2]}{[2+at^2]t} dt} d\tau\right\} \mathrm{e}^{\frac{2k-d}{d-k}\int_c^1 \frac{(-2)[1+at^2]}{[2+at^2]t} dt} dc. \qquad \square$$

The expected uniform price is shown in Figure 4.23 depending on demand and the degree of affiliation. It is easy to see that the uniform price increases with demand as Corollary 4.9 (b) states. Figure 4.24 shows the statement of Corollary 4.9 (a) in the example. The expected uniform prices are higher than the corresponding expected production costs.

The relative difference in the expected uniform price for different degrees of affiliation compared to independent distribution is presented in Figure 4.25. The graph shows that the uniform price of the example can increase as well as decrease with stronger affiliation depending on the demand. Higher as well as lower uniform prices for the strong affiliation can be found for demands up to about $1.175k$ in the example. For greater demands, the expected uniform price is lower than or equal to the price of independent distribution. The greatest decline is only about 2.2%. Hence, consumers cannot expect to pay much less if marginal costs become stronger affiliated.

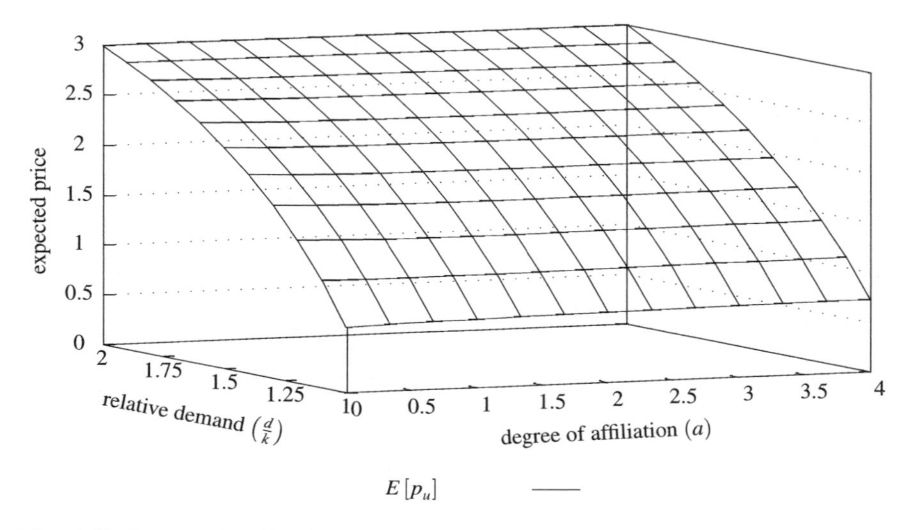

Fig. 4.23 Expected uniform price (case B)

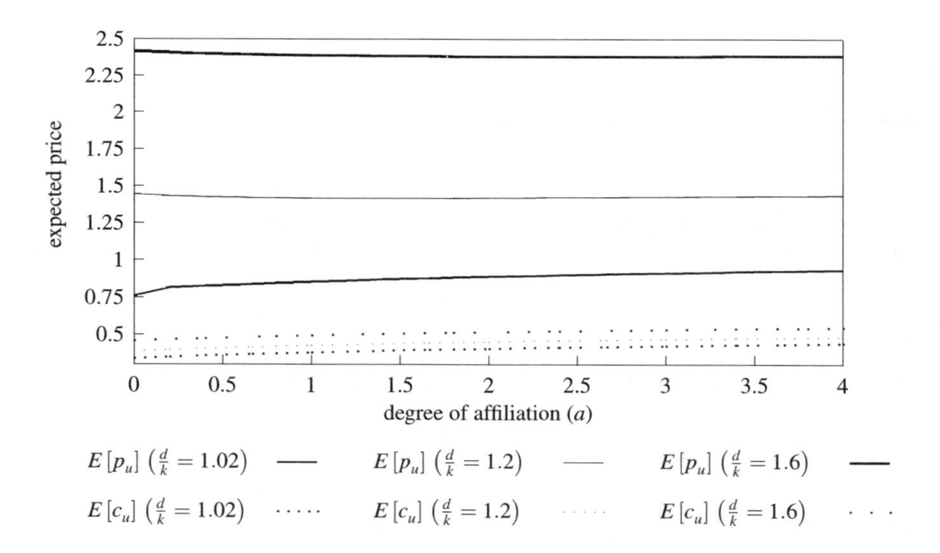

Fig. 4.24 Expected uniform prices for various demands (case B)

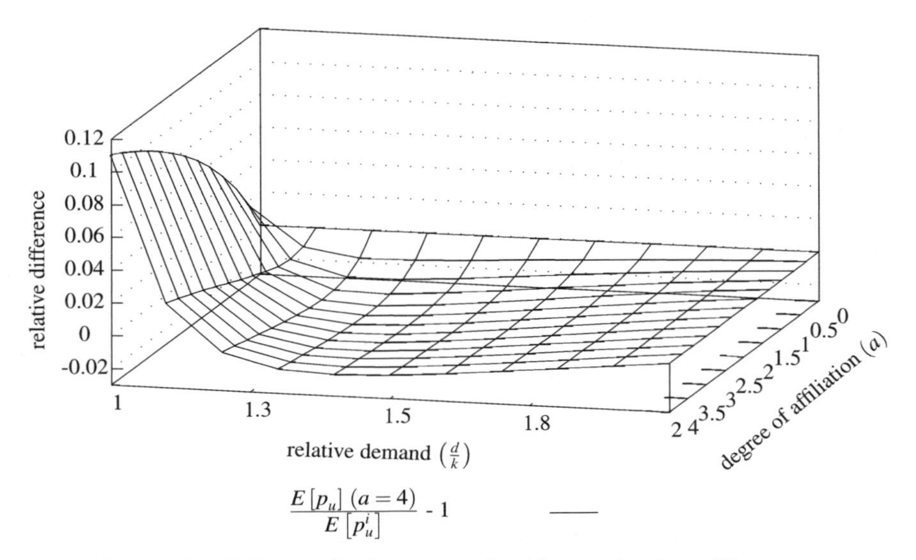

Fig. 4.25 Relative difference in the expected uniform price (case B)

4.4.3 Case C: Full Market Capacity Utilization

If demand is equal to or higher than the aggregated production capacity in the market, consumers can no longer play games with generators. Consequently, the situation is the same as for the discriminatory auction. The results are therefore the same.

Theorem 4.19. *Suppose $d \geq 2k$.*

(a) *The unique Bayes-Nash equilibrium is given by $\left(b_d^*, b_d^*\right)$ with following bid function*

$$b_u^*(c) = \bar{p}. \tag{4.205}$$

(b) *The auction leads to production efficiency with the lowest possible production costs.*

(c) *The expected cost per electricity unit is given by*

$$E[c_u] = \bar{c} - \int_{\underline{c}}^{\bar{c}} F(\tau)\, d\tau. \tag{4.206}$$

(d) *The profit is determined by*

$$\pi_u^*(c) = k[\bar{p} - c]. \tag{4.207}$$

(e) *The contribution margin per capacity unit is given by*

$$\pi_u^{u^*}(c) = \bar{p} - c. \tag{4.208}$$

(f) *The expected uniform price is given by*

$$E[p_u] = \bar{p}. \tag{4.209}$$

Proof.

(a) Because $d \geq 2k$ is given, every generator can sell his entire production capacity k. The output is independent of the bid price, as long as the bid is accepted. Hence, only the contribution margin determines the profit because the price paid to each generator is calculated from their bids. Maximizing the profit means maximizing the contribution margin. The maximum contribution margin is reached with the highest accepted bid. This is in accordance with the price cap rule $\bar{p}$. Notice that the bid is independent of marginal costs.

(b) Because requirements of Theorem 4.1 are satisfied due to optimal bidding $\bar{p}$ and demand $d \geq 2k$, the result of Theorem 4.1 (a) applies.

(c) Because all requirements of Theorem 4.1 are satisfied, the optimal bid $\bar{p}$ is the last accepted bid by the auctioneer and demand is greater than or equal to the aggregated production capacity, (4.8) can be used

$$E[c_u] = E[c] = \bar{c} - \int_{\underline{c}}^{\bar{c}} F(\tau)\, d\tau.$$

(d) This follows directly from the model because the whole production capacity k is sold and the price paid is always $\bar{p}$. Hence, the profit is simply

$$\pi_u^*(c) = k[\bar{p} - c].$$

(e) The contribution margin for each electricity unit, using (4.207), is

$$\pi_u^{u^*}(c) = \frac{\pi_u^*(c)}{k} = \bar{p} - c.$$

(f) This is obvious because all generators bid $\bar{p}$. □

Example 4.13. *Suppose the probability distribution given by Definition 3.5.*

(a) *The optimal bid function is given by*

$$b_u^*(c) = \bar{p}. \tag{4.210}$$

(b) *The expected profit of a generator is obtained by*

$$\pi_u^*(c) = k[\bar{p} - c]. \tag{4.211}$$

(c) *The expected contribution margin per capacity unit is given by*

$$\pi_u^{u^*}(c) = \bar{p} - c. \tag{4.212}$$

(d) *The expected uniform price is given by*

$$E[p_u] = \bar{p}. \tag{4.213}$$

Proof.

(a) - **(c)** are easily checked with Theorem 4.19. □

4.5 Generalized Second-Price Auction

4.5.1 Case A: Competition

The payment of the winner is independent of the offered bids. This leads to a bidding strategy which maximizes only the probability of winning. As long as the loser has remaining unused production capacity, everybody follows this strategy because otherwise a generator with the lowest costs would sell less electricity. The pricing rule clearly shows the competitive advantage of low production costs. Another interesting result is that generators honestly reveal their true costs although they are confidential information.

Theorem 4.20. *Suppose $0 < d < 2k$. The unique Bayes–Nash equilibrium is given by $\left(b_g^*, b_g^*\right)$ with the following bid function*

$$b_g^*(c) = c. \tag{4.214}$$

Proof.

The proof is done in four steps. First, the bid function is found which maximizes the expected profit. Second, it is shown the auction really takes place. Then it is proved that a Bayes–Nash equilibrium is obtained by the bid function and that only one Bayes–Nash equilibrium exists. Due to the symmetry of the generators, the proof is reduced to the analysis of generator i.

Step 1: Bid Function

The expected profit of generator i for $j \neq i$ is

$$\pi_i(b_i, c_i) = kE\left[\{\beta_f(b_j - c_i) + \beta(\bar{p} - c_i)\} 1_{\{b_i < b_j\}}\right.$$
$$\left. + \beta(\bar{p} - c_i) 1_{\{b_i > b_j\}} \,\middle|\, C_i = c_i\right]$$
$$= k\beta_f E\left[(b_j - c_i) 1_{\{b_i < b_j\}} \,\middle|\, C_i = c_i\right] + k\beta[\bar{p} - c_i]. \tag{4.215}$$

A symmetric Bayes–Nash equilibrium is sought. Hence, the competitor calculates his own bid using the bid function i.e. $b_j = b(c_j)$. The equation is then

$$\pi_i(b_i, c_i) = k\left\{\beta_f \int_{b^{-1}(b_i)}^{\bar{c}} [b(c_j) - c_i] f(c_j | c_i)\, dc_j + \beta[\bar{p} - c_i]\right\}. \tag{4.216}$$

The first derivative of the expected profit is given by

$$\frac{\partial \pi_i(b_i, c_i)}{\partial b_i} = (-)k\beta_f \frac{b\left(b^{-1}(b_i)\right) - c_i}{b\left(b^{-1}(b_i)\right)'} f\left(b^{-1}(b_i) | c_i\right)$$
$$= (-)k\beta_f [b_i - c_i] \frac{f\left(b^{-1}(b_i) | c_i\right)}{b\left(b^{-1}(b_i)\right)'}. \tag{4.217}$$

The necessary condition of an optimum is $\frac{\partial \pi_i(b_i, c_i)}{\partial b_i} \stackrel{!}{=} 0$. Because generators are symmetric and β_f is positive, see (4.3), the condition is equivalent to

$$[b(c) - c] \frac{f(c|c)}{b(c)'} = 0.$$

As long as $b(c)' > 0$ holds true and the generator with the highest marginal costs bids highest, i.e. $b(\bar{c}) > b(c)\ \forall c < \bar{c}$, a solution of the problem is

$$b(c) = c, \tag{4.218}$$

because $f(c|c)$ is assumed to be always positive. Due to the design of the auction rules, which makes the generator focus only on maximizing the probability of winning, see (4.216), a generator with the highest marginal costs bidding according to (4.218) knows with certainty that he serves the maximum demand only in the case of a standoff. Hence, he has the strongest incentive to deviate from the bid function. That he does not undercut will be shown in step 3.

The first derivative of the bid function is

$$\frac{\partial b(c)}{\partial c} = 1, \tag{4.219}$$

and the second derivative

$$\frac{\partial^2 b(c)}{\partial c^2} = 0. \tag{4.220}$$

One assumption of the derivative was a strictly increasing bid function. Therefore,

$$\frac{\partial b(c)}{\partial c} > 0, \quad \forall c \in [\underline{c}, \bar{c}], \tag{4.221}$$

has to hold true. This condition is obviously fulfilled due to (4.219). Hence, the bid function is indeed strictly increasing with respect to marginal costs.

Bidding maximizes profit if the second derivative of the expected profit is negative. Partial derivation of (4.217) gives

$$\frac{\partial^2 \pi_i(b_i, c_i)}{\partial b_i^2} = \frac{(-)k\beta_f}{b\left(b^{-1}(b_i)\right)'} \left\{ f\left(b^{-1}(b_i) | c_i\right) + \frac{b_i - c_i}{b\left(b^{-1}(b_i)\right)'} \right.$$
$$\left. \times \left[f_j\left(b^{-1}(b_i) | c_i\right) - f\left(b^{-1}(b_i) | c_i\right) \frac{b\left(b^{-1}(b_i)\right)''}{b\left(b^{-1}(b_i)\right)'} \right] \right\}.$$

Bid function (4.218) is used for calculation in equilibrium. Hence, the second derivative of the profit function, using (4.219) and (4.220), is obtained by

$$\frac{\partial^2 \pi_i(b_i, c_i)}{\partial b_i^2} = (-)\frac{k\beta_f}{b(c_i)'}\left\{\frac{b(c_i)-c_i}{b(c_i)'}\left[f_j(c_i|c_i) - f(c_i|c_i)\frac{b(c_i)''}{b(c_i)'}\right]\right.$$

$$\left. + f(c_i|c_i)\right\}$$

$$= (-)k\beta_f f(c_i|c_i). \tag{4.222}$$

This expression is negative due to definition of β_f, see (4.3), and because $f(c|c)$ is always positive. Hence, bidding according to the bid function (4.218) maximizes the profit.

Step 2: Participation of Generators

An equilibrium only exists if all generators have at least a weak incentive to participate in the auction, i.e. every generator must receive a non-negative expected profit

$$\pi(b(c), c) \geq 0. \tag{4.223}$$

Bidding according to bid function (4.218) provides an expected profit of

$$\pi(b(c), c) = k\left\{\beta_f \int_c^{\bar{c}} [\tau - c] f(\tau|c)\, d\tau + \beta[\bar{p} - c]\right\}$$

$$= k\beta_f \left[\bar{c} - c - \int_c^{\bar{c}} F(\tau|c)\, d\tau\right] + k\beta[\bar{p} - c]. \tag{4.224}$$

It is non-negative, because $\beta_f > 0$ and $\beta \geq 0$, see (4.3) and (4.2). Hence, all generators have an incentive to participate in the auction. Only if demand does not exceed the production capacity of a single generator is the generator with the highest marginal cost indifferent about participating in the auction.

Step 3: Existence of a Bayes–Nash Equilibrium

To prove that a Bayes–Nash equilibrium exists, it is necessary to show that a generator bids according to the bid function $b(c)$, while the other generators are using the same bid function. At first, bid deviations within the value range of the bid function, that is $[b(\underline{c}), b(\bar{c})]$, are focused on. Such a bid deviation can be expressed by using the bid function itself but with adjusted marginal costs. Because the bid function is strictly increasing, it is possible to find an $\varepsilon \neq 0$ so that marginal costs $c - \varepsilon$ lead to this bid. Hence, the following has to be valid for a Bayes–Nash equilibrium with $c - \varepsilon \in [\underline{c}, \bar{c}]$

$$\frac{\partial \pi (b(c-\varepsilon), c)}{\partial b(c-\varepsilon)} \gtrless 0, \qquad \text{for} \quad \varepsilon \gtrless 0. \tag{4.225}$$

Using (4.217), (4.218), and (4.219), the first derivative of the expected profit bidding $b(c-\varepsilon)$ is given by the following expression

$$\begin{aligned}\frac{\partial \pi (b(c-\varepsilon), c)}{\partial b(c-\varepsilon)} &= (-)k\beta_f[b(c-\varepsilon) - c]\frac{f\left(b^{-1}(b(c-\varepsilon))\,|\,c\right)}{b\left(b^{-1}(b(c-\varepsilon))\right)'} \\ &= \varepsilon k \beta_f f(c-\varepsilon|\,c).\end{aligned}$$

The sign is the same as for ε because β_f is positive, see (4.3). Hence, (4.225) is valid and bidding differently from bid function (4.218) within its value range does not occur. Otherwise, it could increase profit to undercut the lowest bid $b(\underline{c})$. This is not rational if the profit of undercutting is not larger than the profit using the common bid function. The statement holds true if the following inequation is satisfied for $\varepsilon \geq 0$

$$\pi (b(\underline{c}) - \varepsilon, c) - \pi (b(c), c) \leq 0. \tag{4.226}$$

Bidding below $b(\underline{c})$ makes winning certain. Consequently, the expected profit, adapting (4.216) and using (4.218), (4.3), is given by

$$\begin{aligned}\pi (b(\underline{c}) - \varepsilon, c) &= k\left\{\beta_f \int_{\underline{c}}^{\bar{c}} [b(\tau) - c] f(\tau|\,c)\, d\tau + \beta[\bar{p} - c]\right\} \\ &= k\beta_f \int_{\underline{c}}^{\bar{c}} \tau f(\tau|\,c)\, d\tau + k\beta\bar{p} - k\alpha c \\ &= k\beta_f \left[\bar{c} - c - \int_{\underline{c}}^{\bar{c}} F(\tau|\,c)\, d\tau\right] + k\beta[\bar{p} - c].\end{aligned} \tag{4.227}$$

The extra profit of using different bidding strategies, see (4.224) and (4.227), is obtained by

$$\pi\left(b(\underline{c})-\varepsilon,c\right)-\pi\left(b\left(c\right),c\right)=(-)k\beta_f\int\limits_{\underline{c}}^{c}F\left(\tau|\,c\right)d\tau. \tag{4.228}$$

It is non-positive because β_f is positive, see (4.3). Therefore, condition (4.226) holds true and no generator undercuts.[5] The last unproved strategy is overbidding the highest bid $b\left(\bar{c}\right)$. By following this strategy the generator is certain to lose the auction. Using (4.216) and presuming $\varepsilon > 0$, the profit is obtained by

$$\pi\left(b(\bar{c})+\varepsilon,c\right)=k\beta\left[\bar{p}-c\right]. \tag{4.229}$$

The extra profit of this deviation is, see (4.224) and (4.229)

$$\pi\left(b(\bar{c})+\varepsilon,c\right)-\pi\left(b\left(c\right),c\right)=(-)k\beta_f\left[\bar{c}-c-\int\limits_{c}^{\bar{c}}F\left(\tau|\,c\right)d\tau\right]. \tag{4.230}$$

Obviously, the extra profit is not positive because β_f is positive, see (4.3). Hence, generators do not earn more by overbidding and do not overbid.[6] No deviation from the bid function takes place after all, which means this bidding strategy represents an equilibrium.

---

[5] A stronger statement than (4.226) would be feasible considering the following case. Presume at least one generator produces with the lowest marginal costs and undercuts $b\left(\underline{c}\right)$. Only such a bidder can do so because is extra profit is zero, see (4.228). The set of all such realizable profits differs from the set bidding according to (4.218) only if another generator with lowest marginal costs also exists undercutting $b\left(\underline{c}\right)$. In that case, the winner makes a loss for $k\beta_f$, see (4.215), because both bids are below marginal costs. Similar results for both bidders in the case of a standoff. Only the loser suffers the same low profit compared to the same situation but using bid function (4.218). Hence, all but the loser increase their bids to avoid deficiency in profits. The process stops when the profit equals the earning of going along with (4.218). This point is not reached until bidding according to the bid function. Hence, no undercutting takes place for generators with the lowest marginal costs. It is remarkable that only the payment rule for a standoff drives the generators back to the bid function. Using other rules for that situation could lead to multiple equilibria.

[6] As footnote 5 explained for (4.226), a stronger statement is feasible. Suppose both generators have highest marginal costs and overbid $b\left(\bar{c}\right)$ but do not bid above $\bar{p}$. Presume additionally the first generator bids higher than the second. Hence, the first bidder is the loser and receives the same profit as if both had bid according to bid function (4.218), see (4.215). Only the winner's profit is increased because of the difference of the loser's bid and marginal costs. This result encourages the loser to bid lower than the winner in order to reverse his position. A standoff generates less profit and is not focused on. This process ends when the winner's advantage of overbidding is gone, that is when both bid according to (4.218). Consequently, there exists a single situation, which also forces generators with the highest marginal costs to go along with the bid function.

Step 4: Uniqueness of the Bayes–Nash Equilibrium

Bid function (4.218) has no free parameter, so it determines a unique Bayes–Nash equilibrium. The final expression of the bid function is given by

$$b_g^*(c) = c. \qquad \square$$

The goal of this auction type was to force generators to reveal their true marginal costs. This has obviously been achieved. Because of the truth telling, the generalized second-price auction is a good reference for comparison. But this interesting characteristic may have a price for the consumers.

Example 4.14.

Suppose the probability distribution given by Definition 3.5. The optimal bid function is given by

$$b_g^*(c) = c. \tag{4.231}$$

Proof.

It is easily checked with (4.214). $\square$

The optimal bid function is shown in Figure 4.26. It is assumed that all generators offer the highest possible bid, the reserve price $\bar{p}$, if demand reaches the market production capacity ($d \geq 2k$).

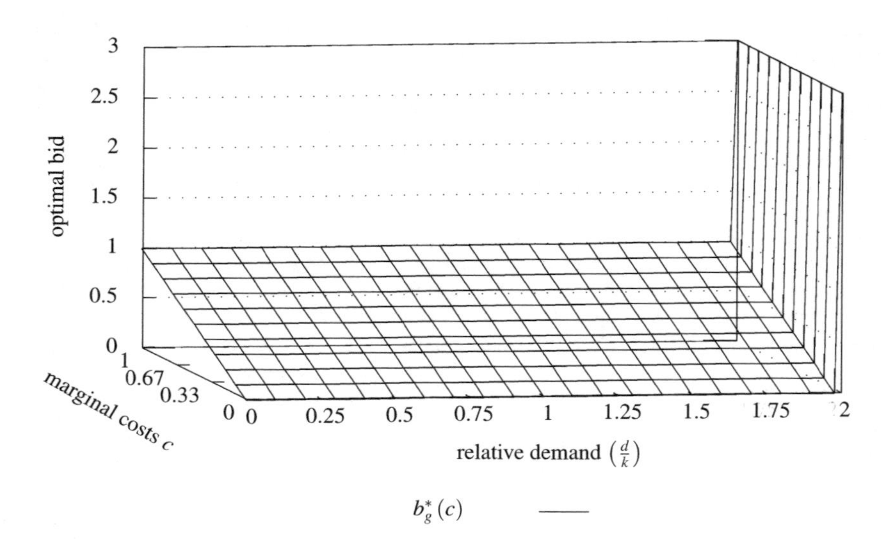

Fig. 4.26 Optimal bid function in a generalized second-price auction

The following theorem relies on the general results of Theorem 4.1.

Theorem 4.21. *Suppose $d \in (0,2k]$.*

(a) *The auction leads to production efficiency. Production costs are the lowest possible.*

(b) *The expected production costs per electricity unit for $d \in (0,k]$ are given by*

$$E\left[c_g\right] = 2\int_{\underline{c}}^{\bar{c}} cF(c)\left[f(c|c) + F_i(c|c)\right]dc - 2\int_{\underline{c}}^{\bar{c}} F(c)\left[1 - F(c|c)\right]dc. \tag{4.232}$$

For independently distributed marginal costs, the expression reduces to

$$E\left[c_g^i\right] = \bar{c} + \int_{\underline{c}}^{\bar{c}} F(c)^2\,dc - 2\int_{\underline{c}}^{\bar{c}} F(c)\,dc. \tag{4.233}$$

If demand is sufficient for both generators, i.e. $d \in (k,2k)$, the production costs are determined by

$$E\left[c_g\right] = 2\left[1 - \frac{k}{d}\right]\bar{c} - 2\frac{k}{d}\int_{\underline{c}}^{\bar{c}} F(c)\,dc + 2\left[\frac{2k}{d} - 1\right] \times \int_{\underline{c}}^{\bar{c}} F(c)\left\{c\left[f(c|c) + F_i(c|c)\right] + F(c|c)\right\}dc. \tag{4.234}$$

The expression for independently distributed marginal costs reduces to

$$E\left[c_g^i\right] = \bar{c} + \left[\frac{2k}{d} - 1\right]\int_{\underline{c}}^{\bar{c}} F(c)^2\,dc - \frac{2k}{d}\int_{\underline{c}}^{\bar{c}} F(c)\,dc. \tag{4.235}$$

Proof.

(a) Optimal bid function (4.214) is increasing in marginal costs and represents bids, which are accepted bids according to the auction rules. Hence, Theorem 4.1 (a) applies.

(b) Based on the same reason given in (a), (4.4) and (4.6) can be used. □

There is no great surprise concerning the production costs. The reason is that the order does not change using bids because bids are equal to the marginal costs. Generators are more interested in the figures given in Theorem 4.22.

Theorem 4.22. *Suppose $d \in (0, 2k]$.*

(a) *The expected profit for $d \in (0, k]$ is given by*

$$\pi_g^*(c) = d\left[\bar{c} - c - \int_c^{\bar{c}} F(\tau|c)\, d\tau\right]. \tag{4.236}$$

The expression for independently distributed marginal costs reduces to

$$\pi_g^{*i}(c) = d\left[\bar{c} - c - \int_c^{\bar{c}} F(\tau)\, d\tau\right]. \tag{4.237}$$

Given $d \in (k, 2k)$, the expected profit is determined by

$$\pi_g^*(c) = [d-k][\bar{p} - \bar{c}] + k[\bar{c} - c] - [2k - d]\int_c^{\bar{c}} F(\tau|c)\, d\tau. \tag{4.238}$$

For independently distributed marginal costs, the expression reduces to

$$\pi_g^{*i}(c) = [d-k][\bar{p} - \bar{c}] + k[\bar{c} - c] - [2k - d]\int_c^{\bar{c}} F(\tau)\, d\tau. \tag{4.239}$$

(b) *The expected contribution margin per capacity unit for $d \in (0, k]$ is obtained by*

$$\pi_g^{u^*}(c) = \frac{d}{k}\left[\bar{c} - c - \int_c^{\bar{c}} F(\tau|c)\, d\tau\right]. \tag{4.240}$$

The expression for independently distributed marginal costs reduces to

$$\pi_g^{u^{*i}}(c) = \frac{d}{k}\left[\bar{c} - c - \int_c^{\bar{c}} F(\tau)\, d\tau\right]. \tag{4.241}$$

The contribution margin for demand $d \in (k, 2k)$ is determined by

$$\pi_g^{u^*}(c) = \left[\frac{d}{k} - 1\right][\bar{p} - \bar{c}] + \bar{c} - c - \left[2 - \frac{d}{k}\right]\int_c^{\bar{c}} F(\tau|c)\, d\tau. \tag{4.242}$$

For independently distributed marginal costs, the expression reduces to

$$\pi_g^{u*i}(c) = \left[\frac{d}{k} - 1\right][\bar{p} - \bar{c}] + \bar{c} - c - \left[2 - \frac{d}{k}\right] \int_c^{\bar{c}} F(\tau)\, d\tau. \tag{4.243}$$

Proof.

(a) The expected profit, using (4.216), (4.214), (4.2), and (4.3), is given by

$$\begin{aligned}
\pi_g^*(c) &= k\left\{\beta_f \int_c^{\bar{c}} [\tau - c] f(\tau|c)\, d\tau + \beta[\bar{p} - c]\right\} \\
&= k\left\{\beta_f [\tau - c] F(\tau|c)\Big|_c^{\bar{c}} - \beta_f \int_c^{\bar{c}} F(\tau|c)\, d\tau + \beta[\bar{p} - c]\right\} \\
&= k\left\{\beta[\bar{p} - c] + \beta_f \left[\bar{c} - c - \int_c^{\bar{c}} F(\tau|c)\, d\tau\right]\right\} \\
&= k\left\{\beta[\bar{p} - \bar{c}] + [\beta_f + \beta][\bar{c} - c] - \beta_f \int_c^{\bar{c}} F(\tau|c)\, d\tau\right\} \\
&= \begin{cases} d\left[\bar{c} - c - \int_c^{\bar{c}} F(\tau|c)\, d\tau\right] & : d \in (0, k], \\ [d - k][\bar{p} - \bar{c}] + k[\bar{c} - c] - [2k - d] \int_c^{\bar{c}} F(\tau|c)\, d\tau & : d \in (k, 2k). \end{cases}
\end{aligned}$$

For independently distributed marginal costs, the expression, using (3.12), reduces to

$$\pi_g^{*i}(c) = \begin{cases} d\left[\bar{c} - c - \int_c^{\bar{c}} F(\tau)\, d\tau\right] & : d \in (0, k], \\ [d - k][\bar{p} - \bar{c}] + k[\bar{c} - c] - [2k - d] \int_c^{\bar{c}} F(\tau)\, d\tau & : d \in (k, 2k). \end{cases}$$

(b) The expected contribution margin for each electricity unit, using (4.236) and (4.238), is obtained by

$$\pi_g^{u*}(c) = \frac{\pi_g^*(c)}{k}$$

$$= \begin{cases} \frac{d}{k}\left[\bar{c} - c - \int\limits_{c}^{\bar{c}} F(\tau|c)\, d\tau\right] & : d \in (0,k], \\ \left[\frac{d}{k} - 1\right][\bar{p} - \bar{c}] + \bar{c} - c - \left[2 - \frac{d}{k}\right] \int\limits_{c}^{\bar{c}} F(\tau|c)\, d\tau & : d \in (k,2k). \end{cases}$$

For independently distributed marginal costs, the expression, using (4.237) and (4.239), reduces to

$$\pi_g^{u*i}(c) = \frac{\pi_g^{*i}(c)}{k}$$
$$= \begin{cases} \frac{d}{k}\left[\bar{c} - c - \int\limits_{c}^{\bar{c}} F(\tau)\, d\tau\right] & : d \in (0,k], \\ \left[\frac{d}{k} - 1\right][\bar{p} - \bar{c}] + \bar{c} - c - \left[2 - \frac{d}{k}\right] \int\limits_{c}^{\bar{c}} F(\tau)\, d\tau & : d \in (k,2k). \end{cases}$$

□

Corollary 4.10. *Suppose* $d \in (0,2k)$.

(a) *The expected profit and expected contribution margin per capacity unit are positive except for the generator with marginal cost* $\bar{c}$ *if the demand lies in the range* $(0,k]$*. This generator expects a profit or contribution margin of zero for this case.*

(b) *The expected profit and expected contribution margin per capacity unit increase strictly with demand.*

(c) *Suppose* $d \in (0,k]$*. The expected profit and expected contribution margin per capacity unit decrease strictly with marginal costs if the following condition holds true for all marginal costs*

$$1 - F(c|c) > (-) \int\limits_{c}^{\bar{c}} F_i(\tau|c)\, d\tau. \tag{4.244}$$

The inequation is satisfied given independent distribution. The slope of the expected profit and expected contribution margin is zero for $\bar{c}$.

(d) *Suppose* $d \in (k,2k)$*. The expected profit and expected contribution margin per capacity unit strictly decrease with marginal costs if the following condition holds true for all marginal costs*

$$\frac{k}{2k-d} - F(c|c) > (-) \int\limits_{c}^{\bar{c}} F_i(\tau|c)\, d\tau. \tag{4.245}$$

This condition is satisfied for the independent distribution.

Proof.

(a) The expected profit for $d \in (0,k]$, see (4.236), is zero for $\bar{c}$. It is positive for the remaining marginal costs if the following holds true

$$0 < \int_{c}^{\bar{c}} [1 - F(\tau|c)]\, d\tau \tag{4.246}$$

$$\Longleftrightarrow$$

$$1 > F(\tau|c) \quad \forall c, \tau \in [\underline{c}, \underline{c}).$$

This is satisfied given $c \in [\underline{c}, \bar{c})$ due to the definition of $F(\cdot|\cdot)$, see (3.5). Hence, the expected profit for $d \in (0,k]$ is positive except for the generator with the highest marginal cost $\bar{c}$. He expects a profit of zero.

The expected profit for $d \in (k, 2k)$, see (4.238), is positive if the following is valid

$$0 < [d-k][\bar{p} - \bar{c}] + k[\bar{c} - c] - [2k - d] \int_{c}^{\bar{c}} F(\tau|c)\, d\tau$$

$$< [d-k][\bar{p} - c] + [2k - d] \int_{c}^{\bar{c}} [1 - F(\tau|c)]\, d\tau.$$

The first summand is positive because $\bar{p} > \bar{c}$, see Assumption 3.12. The second summand is non-negative due to the arguments given above for the demand range $(0,k]$. Hence, the expected profit for all marginal costs is greater than zero.

The same results apply for the expected contribution margin because the factor $\frac{1}{k}$ is positive.

(b) The first derivative of the expected profit with respect to demand, using (4.236), is given by

$$\frac{\partial \pi_g^*(c)}{\partial d} = \bar{c} - c - \int_{c}^{\bar{c}} F(\tau|c)\, d\tau. \tag{4.247}$$

The derivative is obviously zero for $\bar{c}$. It is greater than zero for $c \in [\underline{c}, \bar{c})$ because of the same argument used to show that (4.246) is valid for $c \in [\underline{c}, \bar{c})$. Hence, the expected profit strictly increases with demand. The exception is the generator with marginal cost $\bar{c}$. His profit is independent of demand.

For the demand range $(k, 2k)$, the first derivative of the expected profit, using (4.238), is obtained by

$$\frac{\partial \pi_g^*(c)}{\partial d} = \bar{p} - \bar{c} + \int_{c}^{\bar{c}} F(\tau|c)\, d\tau. \tag{4.248}$$

The derivative is greater than zero because $\bar{p} > \bar{c}$ due to Assumption 3.12 and the integral is non-negative, see (3.5). Hence, the expected profit increases strictly with demand in the range $(k, 2k)$.

The slope for the expected contribution margin, using (4.240) and (4.242), is obtained by

$$\frac{\partial \pi_g^{u^*}(c)}{\partial d} = \frac{1}{k} \frac{\partial \pi_g^*(c)}{\partial d}. \tag{4.249}$$

The same results apply for the expected contribution margin because the factor $\frac{1}{k}$ is positive.

(c) The first derivative of the expected profit for $d \in (0, k]$ with respect to marginal costs, using (4.236), is given by

$$\frac{\partial \pi_g^*(c)}{\partial c} = (-)d \left[1 - F(c|c) + \int_c^{\bar{c}} F_i(\tau|c)\, d\tau \right]. \tag{4.250}$$

The derivative is zero for $\bar{c}$, see (3.5) and (3.14). It is negative if the following holds true for $c \in [\underline{c}, \bar{c})$

$$1 - F(c|c) > (-) \int_c^{\bar{c}} F_i(\tau|c)\, d\tau.$$

The condition for the independent distribution, using (3.12) and (3.14), reduces to

$$F(c) < 1.$$

This is satisfied for $c \in [\underline{c}, \bar{c})$. Hence, the expected profit strictly decreases with marginal costs for the independent distribution.

Similar arguments are found for the expected contribution margin because the factor $\frac{1}{k}$ is independent of c.

(d) The first derivative of the expected profit for $d \in (k, 2k)$ with respect to marginal costs, using (4.238), is given by

$$\frac{\partial \pi_g^*(c)}{\partial c} = [2k - d] F(c|c) - k - [2k - d] \int_c^{\bar{c}} F_i(\tau|c)\, d\tau. \tag{4.251}$$

The expected profit strictly decreases with marginal costs if the derivative is negative, which is given if the following condition holds true

$$\frac{\partial \pi_g^*(c)}{\partial c} < 0$$

$$\Longleftrightarrow$$

$$\frac{k}{2k-d} - F(c|c) > (-) \int_{c}^{\bar{c}} F_i(\tau|c)\, d\tau.$$

For independently drawn marginal costs, the condition, using (3.14) and (3.12), is obtained by

$$\frac{k}{2k-d} - F(c) > 0$$

$$\Longleftrightarrow$$

$$\frac{k}{2k-d} > 1$$

$$\Longleftrightarrow$$

$$d > k.$$

This statement is true for this demand range. Hence, the expected profit always decreases with marginal costs for the independent distribution. □

Although the generators bid their marginal costs, they expect to earn positive profits. The generator with the highest marginal costs has an expected profit of zero due to Corollary 4.10 (a). He stays in the market because he does not lose money and because he has the chance of a standoff, where he earns positive profit. The non-negative profits show that generators do not leave the market. Hence, consumers are sure that they can buy electricity.

Example 4.15. *Suppose the probability distribution given by Definition 3.5.*

(a) *The expected profit for the demand range $(0,k]$ and for independently distributed marginal costs is determined by*

$$\pi_g^{*i}(c) = d\,\frac{[1-c]^2}{2}. \tag{4.252}$$

For strong affiliation, the expression is given by

$$\pi_g^{*}(c) = d\,\frac{[1-c]^2}{2+ac}\left[1+ac\,\frac{2+c}{3}\right]. \tag{4.253}$$

The expression for demand range $(k,2k)$ and for independently distributed marginal costs is obtained by

$$\pi_g^{*i}(c) = [d-k][\bar{p}-c] + [2k-d]\,\frac{1-c^2}{2}. \tag{4.254}$$

The expected profit for strong affiliation is determined by

$$\pi_g^*(c) = [d-k][\bar{p}-c] + [2k-d]\frac{[1-c]^2}{2+ac}\left[1+ac\frac{2+c}{3}\right]. \tag{4.255}$$

(b) *The expected contribution margin per capacity unit for the demand range $(0,k]$ and for independently distributed marginal costs is given by*

$$\pi_g^{u^*i}(c) = \frac{d}{k}\frac{[1-c]^2}{2}. \tag{4.256}$$

The expression for strong affiliation is determined by

$$\pi_g^{u^*}(c) = \frac{d}{k}\frac{[1-c]^2}{2+ac}\left[1+ac\frac{2+c}{3}\right]. \tag{4.257}$$

The contribution margin for the demand range $(k,2k)$ and for independently distributed marginal costs is obtained by

$$\pi_g^{u^*i}(c) = \left[\frac{d}{k}-1\right][\bar{p}-c] + \left[2-\frac{d}{k}\right]\frac{1-c^2}{2}. \tag{4.258}$$

For strong affiliation, the expression is determined by

$$\pi_g^{u^*}(c) = \left[\frac{d}{k}-1\right][\bar{p}-c] + \left[2-\frac{d}{k}\right]\frac{[1-c]^2}{2+ac}\left[1+ac\frac{2+c}{3}\right]. \tag{4.259}$$

Proof.

(a) The expected profit for $d \in (0,k]$ and for independently distributed marginal costs, using (4.237) and (4.196), is determined by

$$\begin{aligned}\pi_g^*(c) &= d\left[1-c-\frac{1-c^2}{2}\right]\\ &= d\frac{[1-c]^2}{2}.\end{aligned}$$

For strong affiliation, the expression, using (4.236) and (4.196), is given by

$$\begin{aligned}\pi_g^*(c) &= d[1-c]\left\{1-\frac{1+c+\frac{a}{3}[1+c+c^2]c}{2+ac}\right\}\\ &= d\frac{1-c}{2+ac}\left[1-c+ac\frac{2-c-c^2}{3}\right]\\ &= d\frac{[1-c]^2}{2+ac}\left[1+ac\frac{2+c}{3}\right].\end{aligned}$$

Using (4.238) and (4.196), the expected profit for $d \in (k, 2k)$ and for independently distributed marginal costs is determined by

$$\begin{aligned}\pi_g^*(c) &= [d-k][\bar{p}-1] + k[1-c] - [2k-d]\left[1-c-\frac{1-c^2}{2}\right]\\ &= [d-k][\bar{p}-c] + [2k-d]\frac{1-c^2}{2}.\end{aligned}$$

The expression for strong affiliation is obtained by

$$\begin{aligned}\pi_g^*(c) &= [d-k][\bar{p}-1] + k[1-c]\\ &\quad -[2k-d]\frac{1-c}{2+ac}\left\{1+c+\frac{a}{3}[1+c+c^2]c\right\}\\ &= [d-k][\bar{p}-c] + [2k-d]\frac{[1-c]^2}{2+ac}\left[1+ac\frac{2+c}{3}\right].\end{aligned}$$

(b) The expected contribution margin per capacity unit for independently distributed marginal costs, using (4.252) and (4.254), is obtained by

$$\begin{aligned}\pi_g^{*i}(c) &= \frac{\pi_g^{*i}(c)}{k}\\ &= \begin{cases} \dfrac{d}{k}\dfrac{[1-c]^2}{2} & : d \in (0,k],\\ \left[\dfrac{d}{k}-1\right][\bar{p}-c] + \left[2-\dfrac{d}{k}\right]\dfrac{1-c^2}{2} & : d \in (k,2k). \end{cases}\end{aligned}$$

For strong affiliation, the expression, using (4.253) and (4.255), is given by

$$\begin{aligned}\pi_g^*(c) &= \frac{\pi_g^*(c)}{k}\\ &= \begin{cases} \dfrac{d}{k}\dfrac{[1-c]^2}{2+ac}\left[1+ac\dfrac{2+c}{3}\right] & : d \in (0,k],\\ \left[\dfrac{d}{k}-1\right][\bar{p}-c] & : d \in (k,2k).\\ \quad +\left[2-\dfrac{d}{k}\right]\dfrac{[1-c]^2}{2+ac}\left[1+ac\dfrac{2+c}{3}\right] & \end{cases}\end{aligned}$$

□

Figure 4.27 shows the expected contribution margin per capacity unit for independently distributed marginal costs. The graph for strong affiliation with $a = 4$ is presented in Figure 4.28. As expected, the contribution margin increases with marginal costs and with demand.

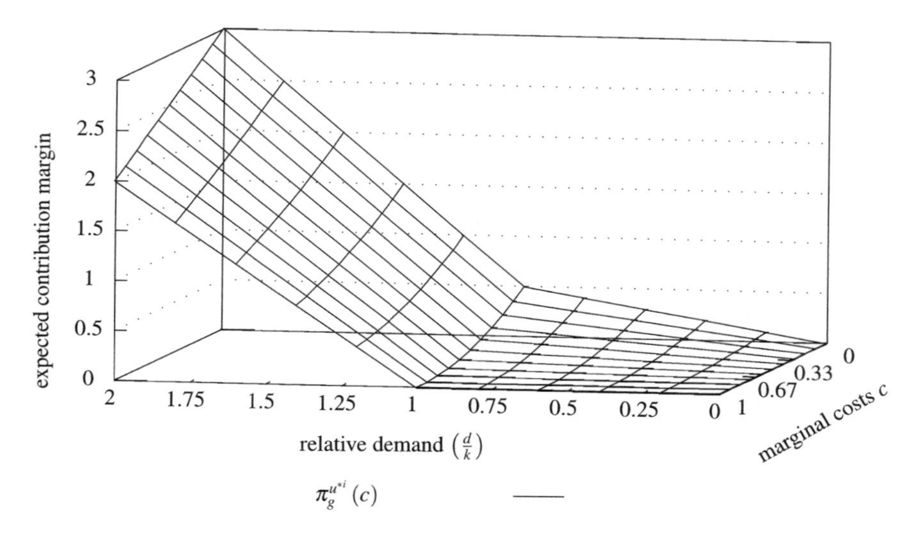

Fig. 4.27 Expected contribution margin in a generalized second-price auction for independently distributed costs

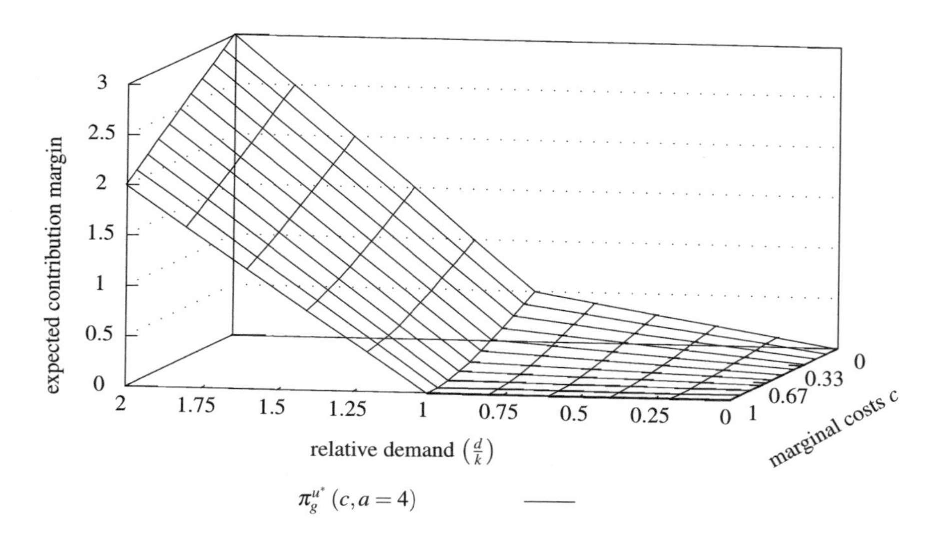

Fig. 4.28 Expected contribution margin in a generalized second-price auction for strong affiliation

The difference in the expected contribution margin for strong affiliation minus independent distribution are shown in Figure 4.29. There are no negative values, i.e. generators expect to earn the same or more if marginal costs become more strongly affiliated. The highest expected excess contribution margin is about 0.05.

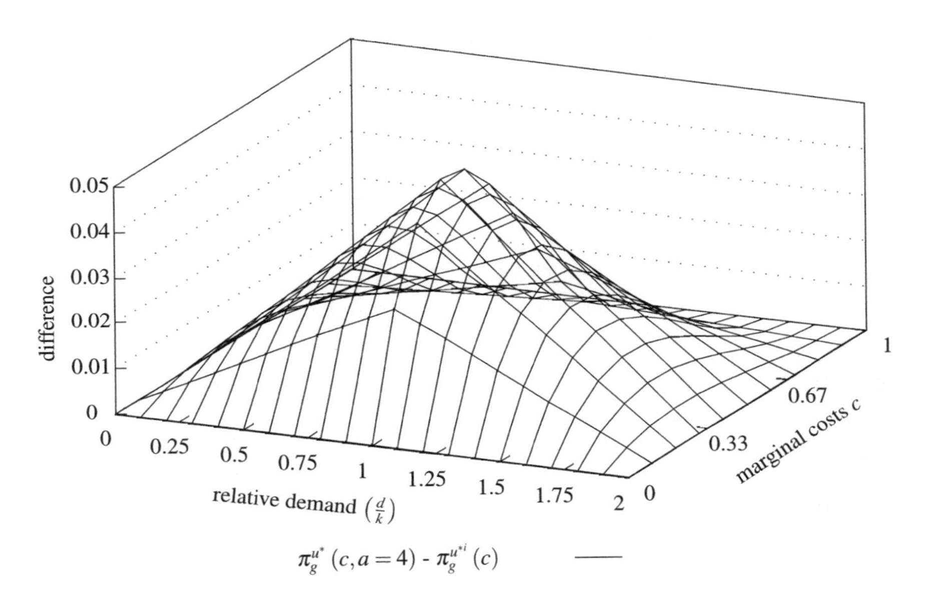

Fig. 4.29 Difference in contribution margins in a generalized second-price auction

The relative difference is more impressive in this case. The graph is shown in Figure 4.30. The highest profit increase is 60% in the graph. Obviously, such high values are only reached for a demand up to k, i.e. the influence of the affiliation among marginal costs pays off most when the profit depends only on the winning case.

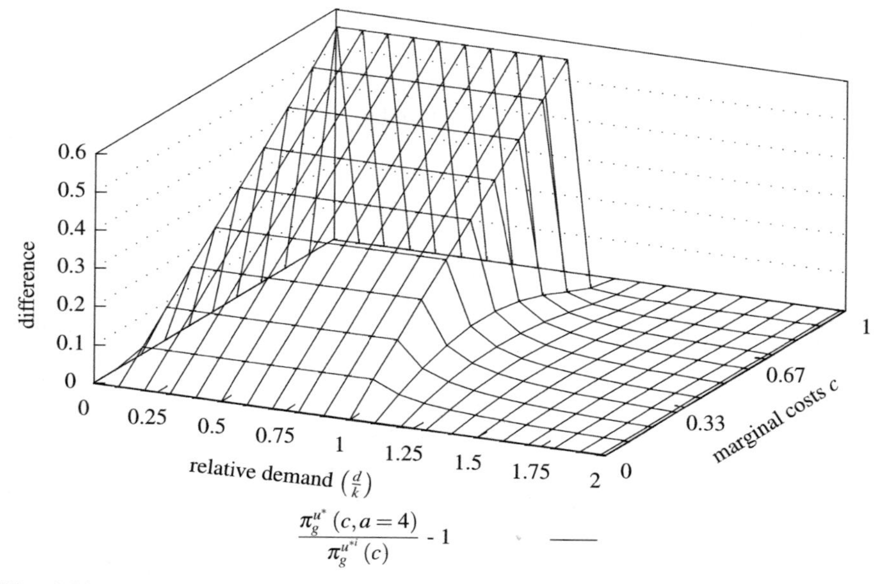

Fig. 4.30 Relative difference in the contribution margins in a generalized second-price auction (case B)

Theorem 4.23. *The expected average price for $d \in (0,k]$ is given by*

$$E\left[p_g\right] = 2\int_{\underline{c}}^{\bar{c}} F(c)\left[cf(c|c) + cF_i(c|c) + \int_{c}^{\bar{c}} F_i(\tau|c)\, d\tau\right] dc. \tag{4.260}$$

The expression for independently distributed marginal costs reduces to

$$E\left[p_g^i\right] = \bar{c} - \int_{\underline{c}}^{\bar{c}} F(c)^2\, dc. \tag{4.261}$$

The expected average price for $d \in (k, 2k)$ is determined by

$$E\left[p_g\right] = 2\left[1 - \frac{k}{d}\right]\bar{p} + 2\left[\frac{2k}{d} - 1\right]\int_{\underline{c}}^{\bar{c}} F(c)\left[cf(c|c) + cF_i(c|c) + \int_{c}^{\bar{c}} F_i(\tau|c)\, d\tau\right] dc. \tag{4.262}$$

For independently distributed marginal costs, the expression reduces to

$$E\left[p_g^i\right] = \left[\frac{2k}{d} - 1\right]\left[\bar{c} - \int\limits_{\underline{c}}^{\bar{c}} F(c)^2\, dc\right] + 2\left[1 - \frac{k}{d}\right]\bar{p}. \tag{4.263}$$

Proof.

The expected average price paid for a unit of electricity, with $j \neq i$, is given by

$$\begin{aligned} E\left[p_g\right] &= \frac{k}{d}E\left[\left\{\beta_f b_g^*(c_j) + \beta\bar{p}\right\} 1_{\left\{b_g^*(c_i) < b_g^*(c_j)\right\}} + \beta\bar{p}\, 1_{\left\{b_g^*(c_i) > b_g^*(c_j)\right\}}\right] \\ &\quad + \frac{k}{d}E\left[\left\{\beta_f b_g^*(c_i) + \beta\bar{p}\right\} 1_{\left\{b_g^*(c_j) < b_g^*(c_i)\right\}} + \beta\bar{p}\, 1_{\left\{b_g^*(c_j) > b_g^*(c_i)\right\}}\right]. \end{aligned}$$

Due to identical bid functions, which are strictly increasing in c, see (4.221), and because information variables are drawn from the same distribution, the expected average price is obtained by

$$\begin{aligned} E\left[p_g\right] &= \frac{2k}{d}E\left[\left\{\beta_f b_g^*(c_j) + \beta\bar{p}\right\} 1_{\left\{b_g^*(c_i) < b_g^*(c_j)\right\}} + \beta\bar{p}\, 1_{\left\{b_g^*(c_i) > b_g^*(c_j)\right\}}\right] \\ &= \frac{2k}{d}E\left[\left\{\beta_f b_g^*(c_j) + \beta\bar{p}\right\} 1_{\left\{c_i < c_j\right\}} + \beta\bar{p}\, 1_{\left\{c_i > c_j\right\}}\right] \\ &= \frac{2k}{d}\left[\beta_f \int\limits_{\underline{c}}^{\bar{c}} \int\limits_{c}^{\bar{c}} b_g^*(\tau) f(\tau, c)\, d\tau\, dc + \beta\bar{p}\right]. \end{aligned} \tag{4.264}$$

Bid function (4.214) is used to obtain

$$E\left[p_g\right] = \frac{2k}{d}\left[\beta_f \int\limits_{\underline{c}}^{\bar{c}} f(c) \int\limits_{c}^{\bar{c}} \tau f(\tau|c)\, d\tau\, dc + \beta\bar{p}\right]$$

$$= \frac{2k}{d}\left\{\beta_f \int_{\underline{c}}^{\bar{c}} f(c)\left[\bar{c} - cF(c|c) - \int_{c}^{\bar{c}} F(\tau|c)\,d\tau\right] dc + \beta\bar{p}\right\}$$

$$= \frac{2k}{d}\left\{\beta_f \bar{c} - \beta_f F(c)\left[cF(c|c) + \int_{c}^{\bar{c}} F(\tau|c)\,d\tau\right]\Bigg|_{\underline{c}}^{\bar{c}}\right.$$

$$+\beta_f \int_{\underline{c}}^{\bar{c}} F(c)\left[F(c|c) + cf(c|c) + cF_i(c|c)\right] dc$$

$$\left.+\beta_f \int_{\underline{c}}^{\bar{c}} F(c)\left[\int_{c}^{\bar{c}} F_i(\tau|c)\,d\tau - F(c|c)\right] dc + \beta\bar{p}\right\}$$

$$= \frac{2k}{d}\left\{\beta_f \int_{\underline{c}}^{\bar{c}} F(c)\left[cf(c|c) + cF_i(c|c) + \int_{c}^{\bar{c}} F_i(\tau|c)\,d\tau\right] dc + \beta\bar{p}\right\}.$$

Definitions of β and β_f, see (4.2) and (4.3), are used to get

$$E\left[p_g\right] = \begin{cases} 2\int_{\underline{c}}^{\bar{c}} F(c)\left[cf(c|c) + cF_i(c|c) + \int_{c}^{\bar{c}} F_i(\tau|c)\,d\tau\right] dc & : d \in (0,k], \\ 2\left[1 - \frac{k}{d}\right]\bar{p} + 2\left[\frac{2k}{d} - 1\right]\int_{\underline{c}}^{\bar{c}} F(c)\left[cf(c|c)\right. & : d \in (k,2k). \\ \quad \left. + cF_i(c|c) + \int_{c}^{\bar{c}} F_i(\tau|c)\,d\tau\right] dc & \end{cases}$$

Using (3.11), (3.14), and (4.11), the expression for independently distributed marginal costs is given by

$$E\left[p_g^i\right] = \begin{cases} 2\int_{\underline{c}}^{\bar{c}} cf(c)F(c)\,dc & : d \in (0,k], \\ 2\left[1 - \frac{k}{d}\right]\bar{p} + 2\left[\frac{2k}{d} - 1\right]\int_{\underline{c}}^{\bar{c}} cf(c)F(c)\,dc & : d \in (k,2k) \end{cases}$$

$$= \begin{cases} \bar{c} - \int\limits_{\underline{c}}^{\bar{c}} F(c)^2 \, dc & : d \in (0,k], \\ 2\left[1 - \frac{k}{d}\right]\bar{p} + \left[\frac{2k}{d} - 1\right]\left[\bar{c} - \int\limits_{\underline{c}}^{\bar{c}} F(c)^2 \, dc\right] & : d \in (k, 2k). \end{cases}$$

□

Corollary 4.11. *Suppose $d \in (0, 2k)$.*

(a) *The expected average price is greater than the expected production costs.*
(b) *The expected average price is independent of demand for the demand range $(0,k]$.*

Proof.

(a) Suppose $c_i \geq c_j \, \forall \, c_i, c_j \in [\underline{c}, \bar{c}]$. The realized average price, using (4.1)–(4.3), is determined by

$$p_g = \frac{\beta_f}{\alpha + \beta} c_i + \frac{2\beta}{\alpha + \beta} \bar{p}$$
$$= \frac{\alpha - \beta}{\alpha + \beta} c_i + \frac{2\beta}{\alpha + \beta} \bar{p}.$$

The realized production costs are obtained by, taking into account that the bids leads to the lowest production costs, see Theorem 4.21 (a)

$$c^r = \frac{\alpha}{\alpha + \beta} c_j + \frac{\beta}{\alpha + \beta} c_i.$$

The difference between the average price and the production costs is given by

$$p_g - c^r = \frac{\alpha - \beta}{\alpha + \beta} c_i + \frac{2\beta}{\alpha + \beta} \bar{p} - \frac{\alpha}{\alpha + \beta} c_j - \frac{\beta}{\alpha + \beta} c_i$$
$$= \frac{1}{\alpha + \beta} \left\{ \alpha [c_i - c_j] + 2\beta [\bar{p} - c_i] \right\}.$$

The expected difference is then the difference between the expected average price minus the expected production costs. It is now determined by

$$E\left[p_g - c^r\right] = \frac{1}{\alpha + \beta} E\left[\left\{ \alpha [c_i - c_j] + 2\beta [\bar{p} - c_i] \right\} 1_{\{c_i \geq c_j\}}\right]$$
$$= \frac{1}{\alpha + \beta} E\left[\left\{ \alpha [c_i - c_j] + 2\beta [\bar{p} - c_i] \right\} 1_{\{c_i > c_j\}}\right].$$

The term $\alpha[c_i - c_j]$ is greater than zero because $c_i > c_j$ and $\alpha > 0$ due to (4.1). The other term $\beta[\bar{p} - c_i]$ is non-negative because $\bar{p} > c_i$ due to Assumption 3.12 and $\alpha \geq 0$ according to (4.2). Hence, the expectation over all combinations of marginal costs is positive because the probability density is positive due to Assumption 3.7. Therefore, the expected average price is greater than the expected production costs.

(b) Suppose $d \in (0,k]$. The first derivative of the expected average price, using (4.264) instead of (4.260), (4.1)–(4.3), and (4.214), is obtained by

$$E\left[p_g\right] \frac{\partial E\left[p_g\right]}{\partial d} = 2 \int_{\underline{c}}^{\bar{c}} \int_{c}^{\bar{c}} \frac{\partial b_g^*(\tau)}{\partial d} f(\tau, c)\, d\tau\, dc$$
$$= 0. \tag{4.265}$$

Hence, the expected average price is independent of demand. □

Consumers must accept that the expected average price is greater than the production costs. It ensures that the generators stay in the market.

Example 4.16.

Suppose the probability distribution given by Definition 3.5. The expected average price for independently distributed marginal costs is obtained by

$$E\left[p_g^i\right] = \begin{cases} \frac{2}{3} & : d \in (0,k] \\ \frac{2}{3}\left[\frac{2k}{d} - 1\right] + 2\left[1 - \frac{k}{d}\right]\bar{p} & : d \in (k,2k). \end{cases} \tag{4.266}$$

For strong affiliated marginal costs, the expression is given by

$$E\left[p_g\right] = \begin{cases} \frac{4}{5} - \frac{8}{15a^3}\left[a^2 - 4a + 16 - \frac{64}{4+a}\right] & : d \in (0,k] \\ 2\left[1 - \frac{k}{d}\right]\bar{p} + \frac{4}{5}\left[\frac{2k}{d} - 1\right] & : d \in (k,2k). \\ \quad \times \left\{1 - \frac{2}{3a^3}\left[a^2 - 4a + 16 - \frac{64}{4+a}\right]\right\} & \end{cases} \tag{4.267}$$

Proof.

The expected average price for independently distributed marginal costs and for $d \in (0,k]$, using (4.261) and (3.25), is obtained by

$$E\left[p_g^i\right] = 1 - \int_0^1 c^2\,dc$$

$$= 1 - \left.\frac{c^3}{3}\right|_0^1 = \frac{2}{3}.$$

The expression for $d \in (k, 2k)$, using (4.263) and previous result, is given by

$$E\left[p_g^i\right] = \frac{2}{3}\left[\frac{2k}{d} - 1\right] + 2\left[1 - \frac{k}{d}\right]\bar{p}.$$

The derivation of the expression for strong affiliation with $a > 0$ uses following term, taking into account (3.25) and (3.32)

$$\int_0^1 F(c) \int_c^1 F_i(\tau|c)\,d\tau\,dc$$

$$= (-)\int_0^1 \frac{1+\frac{a}{4}c}{1+\frac{a}{4}} c \int_c^1 \frac{a[1-\tau]\tau}{2\left[1+\frac{a}{2}c\right]^2}\,d\tau\,dc$$

$$= \frac{(-a)}{3[4+a]}\int_0^1 \frac{4+ac}{[2+ac]^2}[1-3c^2+2c^3]c\,dc$$

$$= \frac{(-1)}{3a^3[4+a]}\int_0^1 [2a^3c^3 - 3a^3c^2 - 8ac + a^3 + 12a + 32]\,dc$$

$$+ \frac{16}{a^3}\frac{2+a}{4+a}\int_0^1 \frac{1}{2+ac}\,dc + \frac{4}{3a^3}\frac{a^3-12a-16}{4+a}\int_0^1 \frac{1}{[2+ac]^2}\,dc$$

$$= (-)\frac{a^3+16a+64}{6a^3[4+a]} + \frac{16}{a^4}\frac{2+a}{4+a}\ln\left(1+\frac{a}{2}\right) + \frac{2}{3a^3}\frac{a^3-12a-16}{[4+a][2+a]}$$

$$= \frac{(-1)}{6a} + \frac{4}{3a^2} - \frac{28}{3a^3} + \frac{64}{3a^3[4+a]} + \frac{16}{a^4}\frac{2+a}{4+a}\ln\left(1+\frac{a}{2}\right). \qquad (4.268)$$

For strong affiliation and $d \in (0, k]$, the expected average price according to (4.260), using (4.17) and (4.268), is given by

$$
\begin{aligned}
E\left[p_g\right] &= \frac{4}{5} - \frac{2}{4+a}\left[\frac{1}{10} + \frac{2}{3a} - \frac{4}{a^2} - \frac{16}{a^3} + 16\frac{2+a}{a^4}\ln\left(1+\frac{a}{2}\right)\right] \\
&\quad -2\left\{\frac{1}{6a} - \frac{4}{3a^2} + \frac{28}{3a^3} - \frac{64}{3a^3[4+a]} - \frac{16[2+a]}{a^4[4+a]}\ln\left(1+\frac{a}{2}\right)\right\} \\
&= \frac{4}{5} - \frac{1}{3a} + \frac{8}{3a^2} - \frac{56}{3a^3} - \frac{2}{4+a}\left[\frac{1}{10} + \frac{2}{3a} - \frac{4}{a^2} - \frac{112}{3a^3}\right] \\
&= \frac{4}{5} - \frac{1}{3a} + \frac{8}{3a^2} - \frac{56}{3a^3} - \frac{2}{30a^3}\left[3a^2 + 8a - 152 - \frac{512}{4+a}\right] \\
&= \frac{4}{5} - \frac{8}{15a^3}\left[a^2 - 4a + 16 - \frac{64}{4+a}\right].
\end{aligned}
$$

Using (4.262) and previous result, the same expression but for $d \in (k, 2k)$ is obtained by

$$
E\left[p_g\right] = 2\left[1 - \frac{k}{d}\right]\bar{p} + \frac{4}{5}\left[\frac{2k}{d} - 1\right]\left\{1 - \frac{2}{3a^3}\left[a^2 - 4a + 16 - \frac{64}{4+a}\right]\right\}. \qquad \square
$$

The expected average price is shown in Figure 4.31. It increases with demand in the range $d \in [k, 2k)$ in the example but not for $d \in (0, k)$. Figure 4.32 shows the expected average price for various demands. It is clear to see that the average price is greater than the production costs.

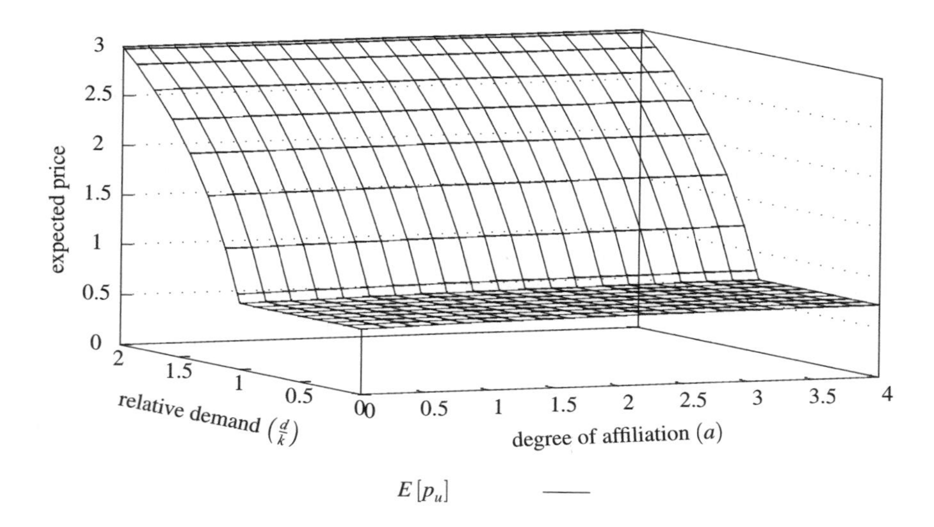

Fig. 4.31 Expected average price in a generalized second-price auction

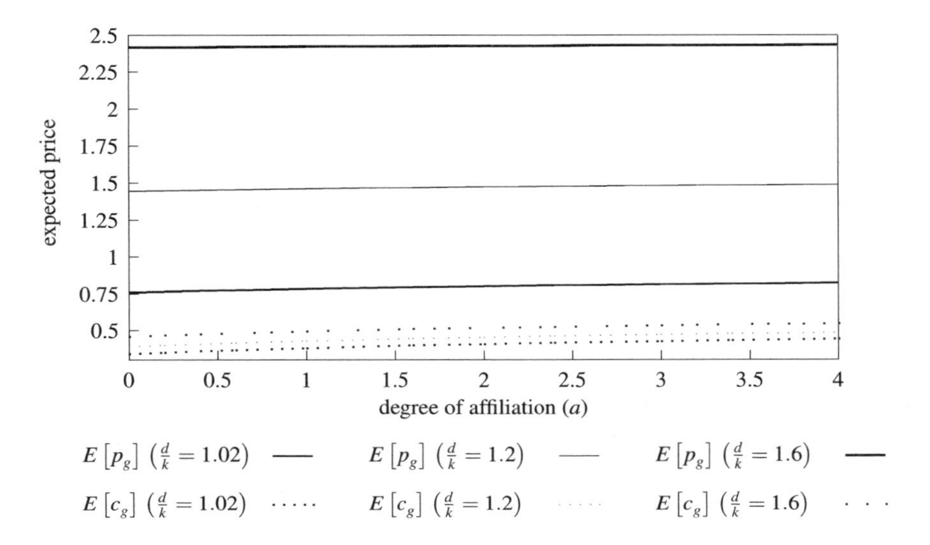

Fig. 4.32 Expected average prices for various demands in a generalized second-price auction (case A)

4.5.2 *Case B: Full Market Capacity Utilization*

If fully utilization is reached, generators no longer stick to the strategy of revealing true costs. No generator is left with unsold electricity. Hence, all bids are rational and are accepted by the auctioneer.

Theorem 4.24. *Suppose $d \geq 2k$.*

(a) *Multiple Bayes–Nash equilibria $\left(b_g^*, b_g^*\right)$ exist with the bid function*

$$b_g^*(c) \in [c, \bar{p}]. \tag{4.269}$$

(b) *The auction leads to production efficiency with the lowest possible production costs.*

(c) *The expected cost per electricity unit is given by*

$$E\left[c_g\right] = \bar{c} - \int_{\underline{c}}^{\bar{c}} F(c)\, dc. \tag{4.270}$$

(d) *The profit is determined by*

$$\pi_g^*(c) = k[\bar{p} - c]. \tag{4.271}$$

(e) *The contribution margin per capacity unit is given by*

$$\pi_g^{u^*}(c) = \bar{p} - c. \tag{4.272}$$

(f) *The expected average price is given by*

$$E\left[p_g\right] = \bar{p}. \tag{4.273}$$

Proof.

(a) Because $d \geq 2k$ is given, every generator can sell his entire production capacity k. The price paid per unit of electricity is independent of the bid of the other bidder, as long as the bid is accepted. Hence, only the participation in the auction is of interest. Therefore, all bids which are accepted bids can be chosen, i.e. $b_g^*(c) \in [c, \bar{p}]$. Consequently, multiple equilibria exist.

(b) Because requirements of Theorem 4.1 are satisfied due to optimal bidding $\bar{p}$ and demand $d \geq 2k$, the result of Theorem 4.1 (a) applies.

(c) Based on the same arguments given in (b), (4.8) can be used to get

$$E\left[c_g\right] = E[c] = \bar{c} - \int_{\underline{c}}^{\bar{c}} F(\tau)\, d\tau.$$

(d) This follows directly from the model because the entire production capacity k is sold and the price paid is always $\bar{p}$. Hence, the profit is simply

$$\pi_g^*(c) = k[\bar{p} - c].$$

(e) The contribution margin for each electricity unit, using (4.271), is

$$\pi_g^{u^*}(c) = \frac{\pi_g^*(c)}{k} = \bar{p} - c.$$

(f) Because of the shortage of production capacity, every generator is always paid a price per unit of electricity that is independent of the competitor's bid, due to the auction rules. The price is $\bar{p}$. □

Example 4.17.

Suppose $d \geq 2k$ and the probability distribution given by Definition 3.5.

(a) *Multiple Bayes–Nash equilibria $\left(b_g^*, b_g^*\right)$ exist with the following bid function*

$$b_g^*(c) \in [c, \bar{p}]. \tag{4.274}$$

(b) *The expected profit is determined by*

$$\pi_g^*(c) = k[\bar{p} - c]. \tag{4.275}$$

(c) *The expected contribution margin per capacity unit is given by*

$$\pi_g^{u^*}(c) = \bar{p} - c. \tag{4.276}$$

(d) *The expected average price is obtained by*

$$E\left[p_g\right] = \bar{p}. \tag{4.277}$$

Proof.

(a) - **(d)** are easily checked with Theorem 4.24. □

4.6 Comparison

All three auction types have been analyzed separately. Generators as well as consumers may want to decide which auction type they would prefer. Therefore, a comparison of the key figures for the auctions completes the analysis.

4.6.1 Case A: A Single Generator Serves the Demand

Corollary 4.12. *Suppose $d \in (0,k]$.*

(a) *Optimal bids are the same in the discriminatory auction and the uniform-price auction.*

$$b_u^*(c) = b_d^*(c). \tag{4.278}$$

(b) *Optimal bids in the generalized second-price auction are lower than in the discriminatory auction.*

$$b_g^*(c) < b_d^*(c). \tag{4.279}$$

(c) *Optimal bids in the generalized second-price auction are lower than in the uniform-price auction.*

$$b_g^*(c) < b_u^*(c). \tag{4.280}$$

Proof.

(a) The equivalence is obvious due to the same decision problem of the bidder in both auction types, see (4.124) and (4.18).

(b) Optimal bids in the generalized second-price auction equal marginal costs, see (4.214). Hence, generators of a discriminatory auction always bid higher if $b_d^*(c) > c$ holds true. This has already been proven by Corollary 4.4 (a).

(c) Because bidding marginal costs are the optimal strategy in the generalized second-price auction, see (4.214), the bids in a uniform-price auction are higher if $b_u^*(c) > c$ is valid. This has already been proven by Corollary 4.4 (a). This is applicable because Theorem 4.11 refers to Theorem 4.2, which is the basis of the proof of Corollary 4.4. □

Bids in the discriminatory auction are the same as those in the uniform-price auction because the payment rule is the same for this demand range. The lowest bids are offered in the generalized second-price auction. The reason is that this auction leads to truth-telling about the marginal costs while the bids in the other auctions are greater than the marginal costs.

Theorem 4.25. *Suppose $d \in (0,k]$.*

(a) *Expected profits are equal in the discriminatory auction and the uniform-price auction*

$$\pi_u^*(c) = \pi_d^*(c). \tag{4.281}$$

The same relation applies for the expected contribution margin.

(b) *The discriminatory auction weakly dominates the generalized second-price auction in terms of expected profits*

$$\pi_g^*(c) \leq \pi_d^*(c) \tag{4.282}$$

if the following condition is satisfied

$$\int_c^{\bar{c}} \frac{1-F(\tau|c)}{1-F(c|c)}\, d\tau \leq b_d^*(c) - c. \tag{4.283}$$

This inequation holds true if the following is valid

$$b_d^*(c) \geq \bar{c}. \tag{4.284}$$

The same relation applies also for the expected contribution margin under the same conditions.

(c) *The uniform-price auction weakly dominates the generalized second-price auction in terms of expected profits*

$$\pi_g^*(c) \leq \pi_u^*(c) \tag{4.285}$$

if the following condition is satisfied

$$\int_{c}^{\bar{c}} \frac{1-F(\tau|c)}{1-F(c|c)}\, d\tau \leq b_u^*(c) - c. \tag{4.286}$$

This inequation holds true if the following is valid:

$$b_u^*(c) \geq \bar{c}. \tag{4.287}$$

The same relation applies for the expected contribution margin under the same conditions.

Proof.

(a) The decision problem of a generator in the discriminatory auction is the same as in the uniform-price auction, see proof of Theorem 4.11. Hence, the calculation of expected profits and therefore optimal bid functions are the same, see (4.18) and (4.124). This ultimately means that the same profits are expected.

The expected contribution margins are also the same, because they differ from the expected profit only in the factor $\frac{1}{k}$.

(b) The inequation, using (4.236) and (4.60)

$$d\left[\bar{c} - c - \int_{c}^{\bar{c}} F(\tau|c)\, d\tau\right]$$

$$\leq d[1-F(c|c)]\left[\bar{p} - \bar{c} + \int_{c}^{\bar{c}} e^{\int_{\tau}^{\bar{c}} \frac{f(t|t)}{1-F(t|t)}\, dt}\, d\tau\right] e^{\int_{c}^{\bar{c}} \frac{(-)f(t|t)}{1-F(t|t)}\, dt}$$

$$\Longleftrightarrow$$

$$\int_{c}^{\bar{c}} [1-F(\tau|c)]\, d\tau \leq [1-F(c|c)]\left[\bar{p} - \bar{c} + \int_{c}^{\bar{c}} e^{\int_{\tau}^{\bar{c}} \frac{f(t|t)}{1-F(t|t)}\, dt}\, d\tau\right] e^{\int_{c}^{\bar{c}} \frac{(-)f(t|t)}{1-F(t|t)}\, dt}$$

holds true for $c = \bar{c}$

$$0 \leq 0.$$

Hence, the relation holds true if the following condition for $c \in [\underline{c}, \bar{c})$, using (4.18), is satisfied

$$\int_{c}^{\bar{c}} \frac{1-F(\tau|c)}{1-F(c|c)} d\tau \leq \left[\bar{p} - \bar{c} + \int_{c}^{\bar{c}} e^{\int_{\tau}^{\bar{c}} \frac{f(t|t)}{1-F(t|t)} dt} d\tau \right] e^{\int_{c}^{\bar{c}} \frac{(-)f(t|t)}{1-F(t|t)} dt}$$

$$\leq b_d^*(c) - c. \tag{4.288}$$

This is condition (4.283). A weaker condition is given by

$$\bar{c} - c \leq b_d^*(c) - c$$

$$\Longleftrightarrow$$

$$b_d^*(c) \geq \bar{c}. \tag{4.289}$$

If this inequation holds true, (4.288) is also satisfied, because the following is valid

$$\bar{c} - c > \int_{c}^{\bar{c}} \frac{1-F(\tau|c)}{1-F(c|c)} d\tau$$

$$\Longleftrightarrow$$

$$0 < \int_{c}^{\bar{c}} \frac{F(\tau|c) - F(c|c)}{1-F(c|c)} d\tau$$

$$\Longleftrightarrow$$

$$F(\tau|c) > F(c|c) \qquad \forall \tau > c.$$

This holds true because $F(\tau|c)$ increases with τ due to (3.5). Hence, the discriminatory auction weakly dominates the generalized second-price auction terms of expected profits if (4.288) is satisfied. This condition is always true if (4.289) is a valid statement.

The comparison of the expected contribution margins leads to same results like the comparison of the expected profits because the expected contribution margins are linked with the expected profits by the same factor $\frac{1}{k}$.

(c) Because the expected profit in the uniform-price auction is the same as in the discriminatory auction due to (a), the same relation as for $\pi_d^*(c)$ holds true for $\pi_u^*(c)$ concerning $\pi_g^*(c)$, see (b).

Because the expected contribution margins are for all auctions $\frac{\pi(c)}{k}$, the statements for the expected profits also hold true for the expected contribution margins. □

The decision problem of bidders in the discriminatory auction and the uniform auction are the same if one generator is sufficient to serve the demand. Hence, the equivalence of expected profits is not surprising as (4.281) shows. Whether a generator receives a higher expected profit in the generalized second-price auction depends on the probability distribution as conditions (4.286) and (4.283) show.

Figure 4.33 illustrates the results. Only the comparison of the generalized second-price auction and the discriminatory auction is presented because the discriminatory auction and the uniform-price auction lead to same values, see Theorem 4.25 (a).

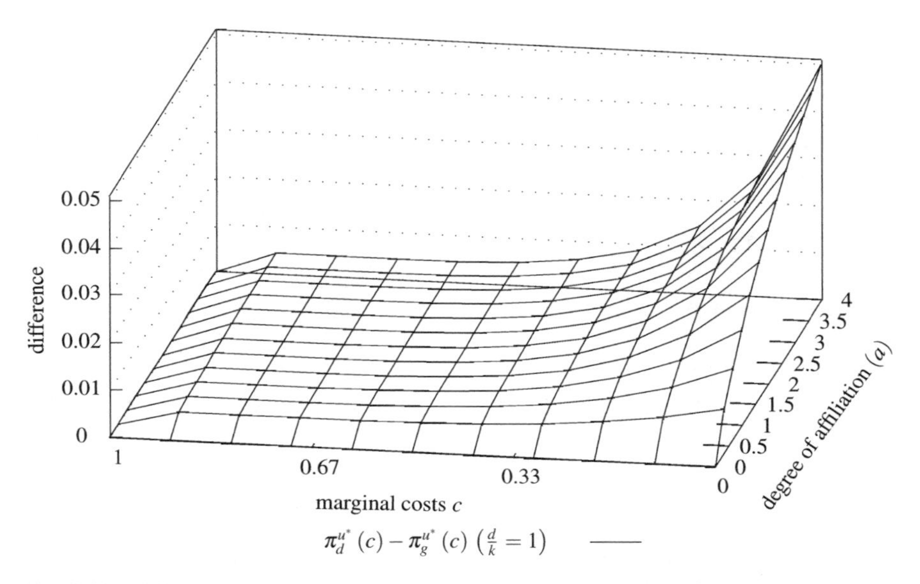

Fig. 4.33 Difference between expected contribution margins of the discriminatory auction and the generalized second-price auction (case A)

The example shows in Figure 4.33 that generators earn higher or same expected profits in the discriminatory auction and in the uniform price auction. Hence, generators do not prefer the generalized second-price auction. They are indifferent between the discriminatory auction and in the uniform-price auction.

Theorem 4.26. *Suppose* $d \in (0,k]$.

(a) *Expected costs per electricity unit are the same for all auction types*

$$E[c_d] = E[c_u] = E\left[c_g\right]. \tag{4.290}$$

(b) *Expected average price in the discriminatory auction equals expected price in the uniform-price auction*

$$E[p_u] = E[p_d]. \tag{4.291}$$

(c) *The expected average price in the discriminatory auction is higher or equal than in the generalized second-price auction*

$$E\left[p_g\right] \leq E\left[p_d\right], \tag{4.292}$$

if the following condition is satisfied

$$\int_{\underline{c}}^{\bar{c}} F(c) \int_{c}^{\bar{c}} F_i(\tau|c)\, d\tau\, dc \leq \int_{\underline{c}}^{\bar{c}} F(c) F_i(c|c) \left[b_d^*(c) - c\right] dc. \tag{4.293}$$

(d) *The expected price in the uniform-price auction is higher or equal than the expected average price in the generalized second-price auction*

$$E\left[p_g\right] \leq E\left[p_u\right], \tag{4.294}$$

if the following condition is satisfied

$$\int_{\underline{c}}^{\bar{c}} F(c) \int_{c}^{\bar{c}} F_i(\tau|c)\, d\tau\, dc \leq \int_{\underline{c}}^{\bar{c}} F(c) F_i(c|c) \left[b_u^*(c) - c\right] dc. \tag{4.295}$$

Proof.

(a) This is obvious because all auction types select the production schedule with the lowest costs given the same demand.

(b) The winning generator in the discriminatory auction as well in the uniform-price auction sets the price for one unit of electricity. Because the optimal bid functions are the same, see (4.18) and (4.124), this yields the same expected average price or uniform price respectively.

(c) The expected average price in the discriminatory auction is higher or equal than in the generalized second-price auction if the following holds true, using (4.260), (4.73), and (4.18)

$$2\int_{\underline{c}}^{\bar{c}} F(c) \left[cf(c|c) + cF_i(c|c) + \int_{c}^{\bar{c}} F_i(\tau|c)\, d\tau \right] dc$$

$$\leq 2\int_{\underline{c}}^{\bar{c}} cF(c) \left[f(c|c) + F_i(c|c)\right] dc + 2\int_{\underline{c}}^{\bar{c}} F(c) F_i(c|c)$$

$$\times \left[\bar{p} - \bar{c} + \int_{c}^{\bar{c}} \mathrm{e}^{\int_{\tau}^{\bar{c}} \frac{f(t|t)}{1-F(t|t)}\, dt}\, d\tau \right] \mathrm{e}^{\int_{c}^{\bar{c}} \frac{(-)f(t|t)}{1-F(t|t)}\, dt}\, dc$$

$$\Longleftrightarrow$$

$$\int_{\underline{c}}^{\bar{c}} F(c) \int_{c}^{\bar{c}} F_i(\tau|c)\, d\tau\, dc \leq \int_{\underline{c}}^{\bar{c}} F(c) F_i(c|c) \left[b_d^*(c) - c\right] dc. \tag{4.296}$$

This is condition (4.293).

(d) Because the expected uniform price in the uniform-price auction is the same as the expected average price in the discriminatory auction due to (a), the same relationship for $E\left[p_d\right]$ concerning $E\left[p_g\right]$ holds true for $E\left[p_u\right]$, see (b). □

All three auction types induce bidding strategies which strongly favor low-cost over high-cost generators. Hence, all auctions select the same production schedule with the lowest costs, which is expressed by Theorem 4.26 (a).

The payments of consumers for electricity purchased can be separated into production costs and profits. Because the first fraction is the same for all auction types, due to Theorem 4.26 (a), the comparison of expected average prices is reduced to the question of earned profits. Because the discriminatory auction leads according to Theorem 4.25 (a) to the same expected profits, it does not surprise that the expected average price equals the expected uniform price, see Theorem 4.26 (b).

The comparison with the generalized second-price auction is more difficult. The reason is that the expected profit of this auction may be higher for some generators but lower for other generators compared to the discriminatory auction or the uniform-price auction, see Theorem 4.26 (c) and (d). Hence, the ranking of the auctions depends on the distribution function.

The difference between the expected average price in the discriminatory auction minus the price of the generalized second-price auction is presented in Figure 4.34. The non-negative values show that the consumers in the discriminatory auction and in the uniform-price auction expect a higher average price per electricity unit than in the generalized second-price auction. This does not surprise because Figure 4.33 showed that the expected profits for all generators in the discriminatory auction and the uniform-price auction are higher or equal than in the generalized second-price auction.

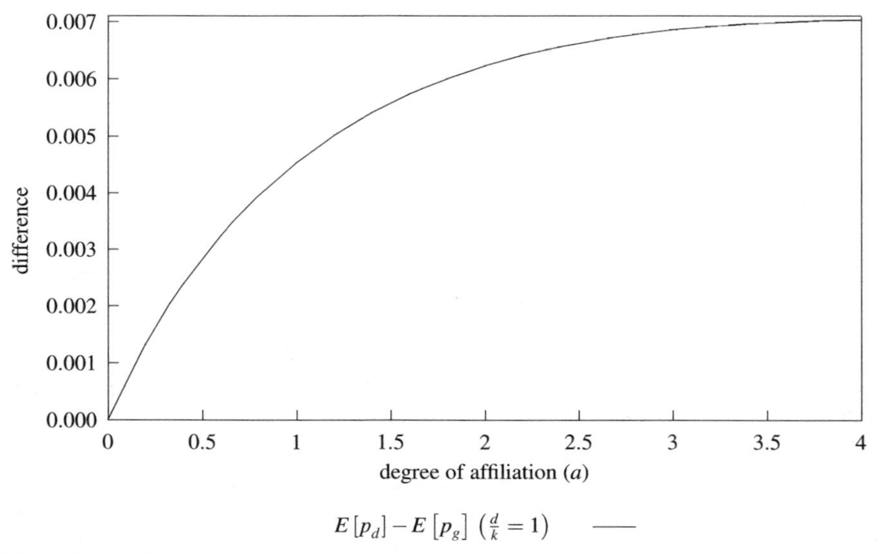

Fig. 4.34 Difference between the expected average price in the discriminatory auction and the generalized second-price auction (case A)

Remark 4.2. *It is easily checked that all auctions have the same properties concerning expected profit, costs, and average price if marginal costs are independently draw.*

4.6.2 Case B: Both Generators Are Necessary

Corollary 4.13. *Suppose $d \in (k, 2k)$ and that the optimal bid functions of all three auction types exist*

(a) *Optimal bids in the discriminatory auction weakly dominate bids in the uniform-price auction*

$$b_u^*(c) < b_d^*(c) \quad \forall c \in [\underline{c}, \bar{c}) \qquad \text{and} \qquad b_u^*(\bar{c}) = b_d^*(\bar{c}). \tag{4.297}$$

(b) *Optimal bids in the generalized second-price auction are lower than in the discriminatory auction*

$$b_g^*(c) < b_d^*(c). \tag{4.298}$$

(c) *Optimal bids in the generalized second-price auction are lower than in the uniform-price auction*

$$b_g^*(c) < b_u^*(c). \tag{4.299}$$

Proof.

(a) Optimal bids in the uniform-price auction are greater than bids in the discriminatory auction if the following holds true for $c < \bar{c}$, see (4.143) and (4.82)

$$\left[\bar{p}-\bar{c}+\int\limits_c^{\bar{c}} e^{\frac{2k-d}{d-k}\int\limits_\tau^{\bar{c}}\frac{f(t|t)}{F(t|t)}\,dt}\,d\tau\right] e^{(-)\frac{2k-d}{d-k}\int\limits_c^{\bar{c}}\frac{f(t|t)}{F(t|t)}\,dt}$$

$$< \left\{\bar{p}-\bar{c}+\int\limits_c^{\bar{c}} e^{\int\limits_\tau^{\bar{c}}\frac{[2k-d]f(t|t)}{k-[2k-d]F(t|t)}\,dt}\,d\tau\right\} e^{(-)\int\limits_c^{\bar{c}}\frac{[2k-d]f(t|t)}{k-[2k-d]F(t|t)}\,dt} \tag{4.300}$$

$$\Longleftrightarrow$$

$$[\bar{p}-\bar{c}]e^{(-)\frac{2k-d}{d-k}\int\limits_c^{\bar{c}}\frac{f(t|t)}{F(t|t)}\,dt}+\int\limits_c^{\bar{c}} e^{(-)\frac{2k-d}{d-k}\int\limits_c^{\tau}\frac{f(t|t)}{F(t|t)}\,dt}\,d\tau$$

$$< [\bar{p}-\bar{c}]e^{(-)\int\limits_c^{\bar{c}}\frac{[2k-d]f(t|t)}{k-[2k-d]F(t|t)}\,dt}+\int\limits_c^{\bar{c}} e^{(-)\int\limits_c^{\tau}\frac{[2k-d]f(t|t)}{k-[2k-d]F(t|t)}\,dt}\,d\tau.$$

Firstly, it is easy to see that the left term equals the right term in case of $c=\bar{c}$, which refers to a generator with the highest costs. He will always bid $\bar{p}$. To prove that all other generators bid differently in both auctions, it is sufficient to show that the following relations are valid for $c \in [\underline{c},\bar{c})$

$$[\bar{p}-\bar{c}]e^{(-)\frac{2k-d}{d-k}\int\limits_c^{\bar{c}}\frac{f(t|t)}{F(t|t)}\,dt} < [\bar{p}-\bar{c}]e^{(-)\int\limits_c^{\bar{c}}\frac{[2k-d]f(t|t)}{k-[2k-d]F(t|t)}\,dt} \tag{4.301}$$

$$\Longleftrightarrow$$

$$e^{(-)\frac{2k-d}{d-k}\int\limits_c^{\bar{c}}\frac{f(t|t)}{F(t|t)}\,dt} < e^{(-)\int\limits_c^{\bar{c}}\frac{[2k-d]f(t|t)}{k-[2k-d]F(t|t)}\,dt}$$

and

$$\int\limits_c^{\bar{c}} e^{(-)\frac{2k-d}{d-k}\int\limits_c^{\tau}\frac{f(t|t)}{F(t|t)}\,dt}\,d\tau < \int\limits_c^{\bar{c}} e^{(-)\int\limits_c^{\tau}\frac{[2k-d]f(t|t)}{k-[2k-d]F(t|t)}\,dt}\,d\tau \tag{4.302}$$

$$\Longleftrightarrow$$

$$e^{(-)\frac{2k-d}{d-k}\int\limits_c^{\tau}\frac{f(t|t)}{F(t|t)}\,dt} < e^{(-)\int\limits_c^{\tau}\frac{[2k-d]f(t|t)}{k-[2k-d]F(t|t)}\,dt} \qquad \forall \tau \in (c,\bar{c}].$$

The validity of (4.301) and (4.302) is given for $c \in [\underline{c},\bar{c})$ if the following holds true

$$(-)\frac{2k-d}{d-k}\frac{f(t|t)}{F(t|t)} < (-)\frac{[2k-d]f(t|t)}{k-[2k-d]F(t|t)} \qquad \forall t \in [\underline{c},\bar{c}) \tag{4.303}$$

$$\Longleftrightarrow$$

$$[d-k]F(c|c) < k-[2k-d]F(c|c)$$

$$\Longleftrightarrow$$

$$k[1-F(c|c)] > 0$$

$$\Longleftrightarrow$$

$$F(c|c) < 1.$$

The last inequation is valid here because the distribution function $F(c|c)$ is strictly increasing in c and reaches its maximum 1 at $\bar{c}$. The maximum was excluded from the analysis, hence (4.303) and consequently (4.301), (4.302), and (4.300) hold true.

(b) Generators in the generalized second-price auction bid their marginal costs, see (4.214). Hence, optimal bids in a discriminatory auction are always higher if $b_d^*(c) > c$ holds true. This has already been proven, see Corollary 4.4 (a).

(c) Because bidding marginal costs are the optimal strategy in the generalized second-price auction, see (4.214), the bids in a uniform-price auction are higher if $b_u^*(c) > c$ is valid. The proof has already been done, see (4.158). □

That bids in the generalized second-price auction are the lowest is not surprising because they equal marginal costs. Bids in the discriminatory auction are greater than those in the uniform-price auction because the generators have different incentives to balance two goals. The first is increasing the probability of winning by bidding low. The second is increasing the price paid by bidding high. Both are counterparts.

A bidder in the discriminatory auction cannot rely on a higher bid by his competitor to increase his own price. In the uniform-price auction, there exists an incentive to bid very low to serve the maximum demand while the competitor sets the price. This price is not known before bidding, but the generator knows that the price paid will not be less than his bid.

Theorem 4.27. *Suppose $d \in (k, 2k)$ and that the optimal bid functions of all three auction types exist.*

(a) *The discriminatory auction weakly dominates the uniform-price auction in terms of expected profits, i.e.*

$$\pi_u^*(c) \le \pi_d^*(c), \tag{4.304}$$

if the following condition is satisfied

$$\int_{c}^{\bar{c}} b_u^*(\tau) f(\tau|c)\, d\tau$$

$$\leq [1 - F(c|c)] b_d^*(c) + \left[\frac{d}{k} - 1\right] F(c|c) [b_d^*(c) - b_u^*(c)]. \qquad (4.305)$$

The same relation applies for the expected contribution margin under the same conditions.

(b) *The discriminatory auction weakly dominates the generalized second-price auction in terms of expected profits*

$$\pi_g^*(c) \leq \pi_d^*(c), \qquad (4.306)$$

if the following condition is satisfied

$$\frac{[d-k]\bar{p} + [2k-d]\left[\bar{c} - F(c|c)c - \int_{c}^{\bar{c}} F(\tau|c)\, d\tau\right]}{k - [2k-d] F(c|c)} \leq b_d^*(c). \qquad (4.307)$$

The same relation applies for the expected contribution margin under the same conditions.

(c) *The uniform-price auction weakly dominates the generalized second-price auction in terms of expected profits*

$$\pi_g^*(c) \leq \pi_u^*(c), \qquad (4.308)$$

if the following condition is satisfied

$$\int_{c}^{\bar{c}} b_u^*(\tau) f(\tau|c)\, d\tau \geq \left[\frac{d}{k} - 1\right] [\bar{p} - F(c|c) b_u^*(c)]$$

$$+ \left[2 - \frac{d}{k}\right] \left[\bar{c} - F(c|c)c - \int_{c}^{\bar{c}} F(\tau|c)\, d\tau\right]. \qquad (4.309)$$

The same relation applies for the expected contribution margin under the same conditions.

Proof.

(a) The expected profit in a discriminatory auction is not lower than in a uniform price auction, if the following condition holds true, taking into account the profit functions, see (4.98) and (4.183), as well as the bid functions, see (4.82) and (4.143)

$$
\begin{aligned}
0 &\leq \pi_d^*(c) - \pi_u^*(c) \\
&\leq \{k - [2k-d]F(c|c)\}\left\{\bar{p} - \bar{c} + \int_c^{\bar{c}} \mathrm{e}^{\int_\tau^{\bar{c}} \frac{[2k-d]f(t|t)}{k-[2k-d]F(t|t)}\,dt}\, d\tau\right\} \mathrm{e}^{\int_c^{\bar{c}} \frac{(-)[2k-d]f(t|t)}{k-[2k-d]F(t|t)}\,dt} \\
&\quad -[d-k]F(c|c)\left[\bar{p} - \bar{c} + \int_c^{\bar{c}} \mathrm{e}^{\frac{2k-d}{d-k}\int_\tau^{\bar{c}} \frac{f(t|t)}{F(t|t)}\,dt}\, d\tau\right] \mathrm{e}^{\frac{2k-d}{d-k}\int_c^{\bar{c}} \frac{(-)f(t|t)}{F(t|t)}\,dt} \\
&\quad -k\int_c^{\bar{c}} f(\xi|c)\left[\bar{p} - \bar{c} + \int_\xi^{\bar{c}} \mathrm{e}^{\frac{2k-d}{d-k}\int_\tau^{\bar{c}} \frac{f(t|t)}{F(t|t)}\,dt}\, d\tau\right] \mathrm{e}^{\frac{2k-d}{d-k}\int_\xi^{\bar{c}} \frac{(-)f(t|t)}{F(t|t)}\,dt}\, d\xi \\
&\quad -k\int_c^{\bar{c}} [1 - F(\tau|c)]\, d\tau \\
&\leq \{k - [2k-d]F(c|c)\}[b_d^*(c) - c] - [d-k]F(c|c)[b_u^*(c) - c] \\
&\quad -k\int_c^{\bar{c}} [b_u^*(\tau) - \tau]f(\tau|c)\, d\tau - k\int_c^{\bar{c}} [1 - F(\tau|c)]\, d\tau \\
&\leq [d-k]F(c|c)[b_d^*(c) - b_u^*(c)] + k\int_c^{\bar{c}} [b_d^*(c) - b_u^*(\tau)]f(\tau|c)\, d\tau \\
&\quad -k\int_c^{\bar{c}} [1 - F(\tau|c)]\, d\tau + k\int_c^{\bar{c}} [\tau - c]f(\tau|c)\, d\tau \\
&\leq [d-k]F(c|c)[b_d^*(c) - b_u^*(c)] + k\int_c^{\bar{c}} [b_d^*(c) - b_u^*(\tau)]f(\tau|c)\, d\tau \\
&\quad -k[\bar{c} - c] + k\,[\tau - c]F(\tau|c)\Big|_c^{\bar{c}} \\
&\leq [d-k]F(c|c)[b_d^*(c) - b_u^*(c)] + k[1 - F(c|c)]b_d^*(c) \\
&\quad -k\int_c^{\bar{c}} b_u^*(\tau)\, f(\tau|c)\, d\tau
\end{aligned}
$$

$\Longleftrightarrow$

$$\int_{c}^{\bar{c}} b_u^*(\tau) f(\tau|c)\, d\tau \leq [1 - F(c|c)] b_d^*(c) + \left[\frac{d}{k} - 1\right] F(c|c) \left[b_d^*(c) - b_u^*(c)\right].$$

This is condition (4.305).

Because the expected contribution margin is for all auctions $\frac{\pi(c)}{k}$, the relationship for the expected profits also holds true for the expected contribution margins.

(b) The stated relation is valid if the following condition, using (4.238), (4.98), and (4.82), holds true

$$[d-k][\bar{p}-\bar{c}] + k[\bar{c}-c] - [2k-d]\int_{c}^{\bar{c}} F(\tau|c)\, d\tau$$

$$\leq \{k - [2k-d]F(c|c)\}\left\{\bar{p} - \bar{c} + \int_{c}^{\bar{c}} e^{\int_{\tau}^{\bar{c}} \frac{[2k-d]f(t|t)}{k-[2k-d]F(t|t)}\, dt}\, d\tau\right\} e^{\int_{c}^{\bar{c}} \frac{(-)[2k-d]f(t|t)}{k-[2k-d]F(t|t)}\, dt}$$

$$\leq \{k - [2k-d]F(c|c)\}\left[b_d^*(c) - c\right]$$

$$\Longleftrightarrow$$

$$b_d^*(c) \geq \frac{[d-k]\bar{p} + [2k-d]\left[\bar{c} - F(c|c)\, c - \int_{c}^{\bar{c}} F(\tau|c)\, d\tau\right]}{k - [2k-d]F(c|c)}. \tag{4.310}$$

This is condition (4.307).

Because the expected contribution margin is for all auctions $\frac{\pi(c)}{k}$, the relationship for the expected profits also holds true for the expected contribution margins.

(c) The relation is valid if the following condition holds true, using (4.238), (4.183), and (4.143)

$$[d-k][\bar{p}-\bar{c}]+k[\bar{c}-c]-[2k-d]\int_{c}^{\bar{c}} F(\tau|c)\,d\tau$$

$$\leq [d-k]F(c|c)\left[\bar{p}-\bar{c}+\int_{c}^{\bar{c}} \mathrm{e}^{\frac{2k-d}{d-k}\int_{\tau}^{\bar{c}}\frac{f(t|t)}{F(t|t)}\,dt}\,d\tau\right]\mathrm{e}^{\frac{2k-d}{d-k}\int_{c}^{\bar{c}}\frac{(-)f(t|t)}{F(t|t)}\,dt}$$

$$+k\int_{c}^{\bar{c}} f(\xi|c)\left[\bar{p}-\bar{c}+\int_{\xi}^{\bar{c}} \mathrm{e}^{\frac{2k-d}{d-k}\int_{\tau}^{\bar{c}}\frac{f(t|t)}{F(t|t)}\,dt}\,d\tau\right]\mathrm{e}^{\frac{2k-d}{d-k}\int_{\xi}^{\bar{c}}\frac{(-)f(t|t)}{F(t|t)}\,dt}\,d\xi$$

$$+k\int_{c}^{\bar{c}}[1-F(\tau|c)]\,d\tau$$

$$\leq [d-k]F(c|c)\,[b_u^*(c)-c]+k\int_{c}^{\bar{c}} f(\tau|c)\,[b_u^*(\tau)-\tau]\,d\tau$$

$$+k\int_{c}^{\bar{c}}[1-F(\tau|c)]\,d\tau$$

$$\leq [d-k]F(c|c)\,[b_u^*(c)-c]+k\int_{c}^{\bar{c}} b_u^*(\tau)\,f(\tau|c)\,d\tau-k[\bar{c}-F(c|c)\,c]$$

$$+k\int_{c}^{\bar{c}} F(\tau|c)\,d\tau+k\int_{c}^{\bar{c}}[1-F(\tau|c)]\,d\tau$$

$$\leq [d-k]F(c|c)\,b_u^*(c)+k\int_{c}^{\bar{c}} b_u^*(\tau)\,f(\tau|c)\,d\tau-\{k-[2k-d]F(c|c)\}\,c$$

$$\Longleftrightarrow$$

$$\int_{c}^{\bar{c}} b_u^*(\tau)\,f(\tau|c)\,d\tau \geq \left[\frac{d}{k}-1\right][\bar{p}-F(c|c)\,b_u^*(c)]$$

$$+\left[2-\frac{d}{k}\right]\left[\bar{c}-F(c|c)\,c-\int_{c}^{\bar{c}} F(\tau|c)\,d\tau\right]. \quad (4.311)$$

This is condition (4.309).

Because the expected contribution margin is for all auctions $\frac{\pi(c)}{k}$, the relationship for the expected profits also holds true for the expected contribution margins. □

There does not exist an auction that leads always to the highest expected profits. Which auction a generator prefers depends therefore on the probability density function.

Figure 4.35 shows the difference between the expected contribution margins in the discriminatory auction and the uniform-price auction for the strong affiliation with $a = 4$. It is easy to see that the discriminatory auction leads always to higher or same values in the example. Hence, a generator prefers the discriminatory auction.

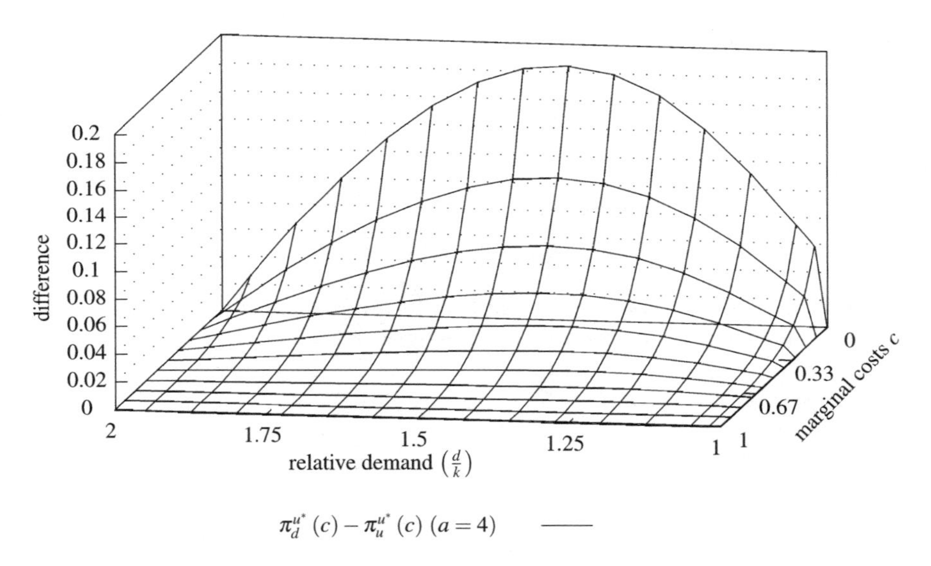

Fig. 4.35 Difference between expected contribution margins of the discriminatory auction and the uniform-price auction (case B)

Figure 4.36 compares the expected contribution margins in the discriminatory auction and the generalized second-price auction. Up to a demand of about $1.15k$, some generators earn higher expected contribution margins in the discriminatory auction and other generators in the generalized second-price auction. Nevertheless, all generators prefer the generalized second-price auction due to higher or same profits if demand is greater than $1.15k$.

Figure 4.37 shows the difference between the expected contribution margins in the generalized second-price auction and the uniform-price auction. Non-negative values mean that the contribution margin is higher or the same in the generalized second-price auction. This is always true in the example for a demand greater than k.

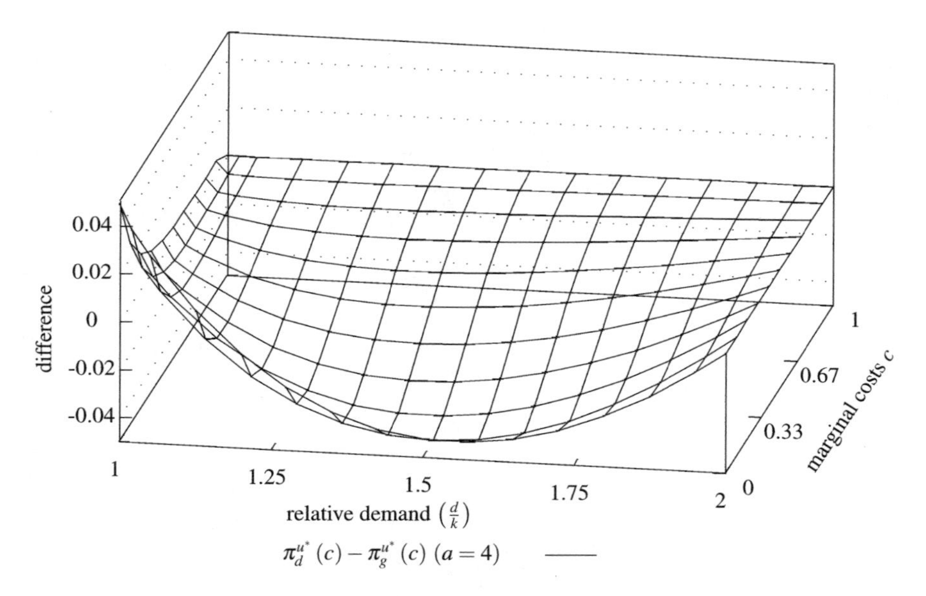

Fig. 4.36 Difference between expected contribution margins of the discriminatory auction and the generalized second-price auction (case B)

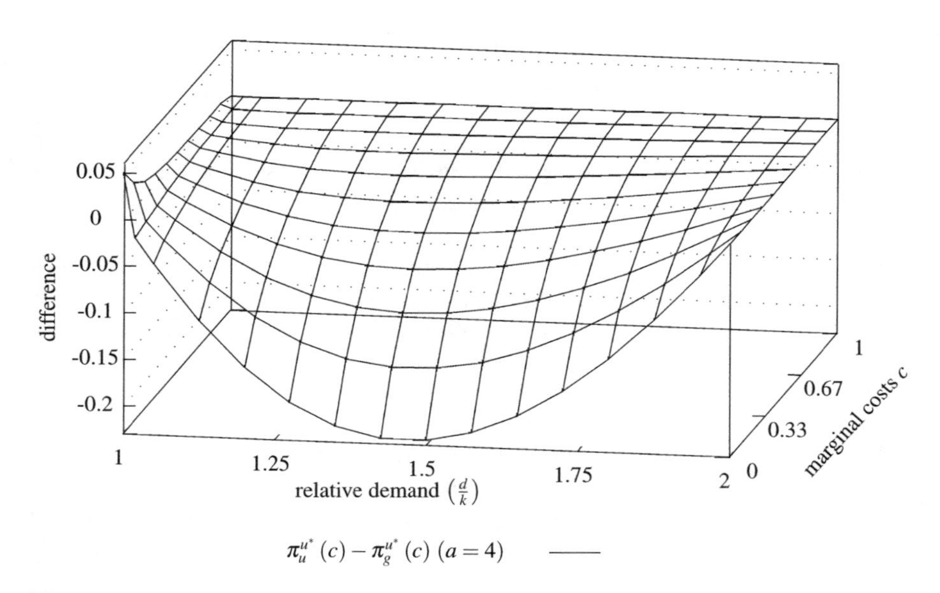

Fig. 4.37 Difference between expected contribution margins of the uniform-price auction and the generalized second-price auction (case B)

Theorem 4.28 compares the expected production costs and the expected average prices.

Theorem 4.28. *Suppose $d \in (k, 2k)$ and that the optimal bid functions of all three auction types exist.*

(a) *The expected costs per electricity unit are the same for all auction types*

$$E\left[c_d\right] = E\left[c_u\right] = E\left[c_g\right]. \tag{4.312}$$

(b) *The expected price of the uniform-price auction is higher than or equal to the expected average price of the discriminatory auction, i.e.*

$$E\left[p_u\right] \geq E\left[p_d\right], \tag{4.313}$$

if the following condition is satisfied

$$\int_{\underline{c}}^{\bar{c}} f(c)\, F(c|c)\left[b_d^*(c) - b_u^*(c)\right] dc \leq \frac{2k}{d} \int_{\underline{c}}^{\bar{c}} f(c)\, F(c|c)\, b_d^*(c)\, dc. \tag{4.314}$$

Otherwise, the relationship is reversed to

$$E\left[p_u\right] < E\left[p_d\right]. \tag{4.315}$$

(c) *The discriminatory auction weakly dominates the generalized second-price auction in terms of expected average prices*

$$E\left[p_g\right] \leq E\left[p_d\right], \tag{4.316}$$

if the following condition is satisfied

$$\int_{\underline{c}}^{\bar{c}} F(c) \int_{c}^{\bar{c}} F_i(\tau|c)\, d\tau\, dc \leq \int_{\underline{c}}^{\bar{c}} F(c)\, F_i(c|c)\left[b_d^*(c) - c\right] dc. \tag{4.317}$$

(d) *The expected price in the uniform-price auction is higher than or equal to the expected average price in the generalized second-price auction*

$$E\left[p_g\right] \leq E\left[p_u\right], \tag{4.318}$$

if the following condition is satisfied

$$\int_{\underline{c}}^{\bar{c}} F(c)\left[\frac{kf(c|c)}{d-k}+F_i(c|c)\right]b_u^*(c)\,dc$$
$$\leq \frac{k}{d}\left[\bar{p}+\frac{2k-d}{d-k}\int_{\underline{c}}^{\bar{c}} cF(c)f(c|c)\,dc\right]$$
$$-\left[\frac{2k}{d}-1\right]\int_{\underline{c}}^{\bar{c}} F(c)\left[F_i(c|c)c+\int_{c}^{\bar{c}} F_i(\tau|c)\,d\tau\right]dc. \qquad (4.319)$$

Proof.

(a) This is obvious because all auction types select the production schedule with the lowest costs given the same demand.

(b) The expected average price in the discriminatory auction is higher than or equal to the expected uniform price if the following condition holds true

$$0 \leq E[p_u]-E[p_d].$$

Using (4.199), (4.75), (4.20), and (4.1)–(4.3), the condition is rewritten to

$$0 \leq 2\int_{\underline{c}}^{\bar{c}} b_u^*(c)f(c)F(c|c)\,dc-2\left[\frac{2k}{d}-1\right]\int_{\underline{c}}^{\bar{c}} b_d^*(c)f(c)F(c|c)\,dc$$

$$\Longleftrightarrow$$

$$0 \leq \frac{2k}{d}\int_{\underline{c}}^{\bar{c}} f(c)F(c|c)b_d^*(c)\,dc-\int_{\underline{c}}^{\bar{c}} f(c)F(c|c)\left[b_d^*(c)-b_u^*(c)\right]dc$$

$$\Longleftrightarrow$$

$$\int_{\underline{c}}^{\bar{c}} f(c)F(c|c)\left[b_d^*(c)-b_u^*(c)\right]dc \leq \frac{2k}{d}\int_{\underline{c}}^{\bar{c}} f(c)F(c|c)b_d^*(c)\,dc.$$

This is condition (4.314). The left term is positive due to (4.297). The right term is also positive because optimal bids are greater than marginal costs according to Corollary 4.4 (a), which are non-negative because of Assumption 3.7.

(c) The expected average price in the discriminatory auction is higher than or equal to the expected average price of the generalized second-price auction if the following holds true, taking into account (4.262), (4.110), and (4.82)

$$2\left[1-\frac{k}{d}\right]\bar{p}+2\left[\frac{2k}{d}-1\right]\int_{\underline{c}}^{\bar{c}} F(c)\left[cf(c|c)+cF_i(c|c)+\int_{c}^{\bar{c}} F_i(\tau|c)\,d\tau\right]dc$$

$$\leq 2\left[1-\frac{k}{d}\right]\bar{p}+2\left[\frac{2k}{d}-1\right]\left\{\int_{\underline{c}}^{\bar{c}} cF(c)\left[f(c|c)+F_i(c|c)\right]dc\right.$$

$$\left.+\int_{\underline{c}}^{\bar{c}} F(c)F_i(c|c)\left[b_d^*(c)-c\right]dc\right\}$$

$$\Longleftrightarrow$$

$$\left[\frac{2k}{d}-1\right]\int_{\underline{c}}^{\bar{c}} F(c)\int_{c}^{\bar{c}} F_i(\tau|c)\,d\tau\,dc$$

$$\leq\left[\frac{2k}{d}-1\right]\int_{\underline{c}}^{\bar{c}} F(c)F_i(c|c)\left[b_d^*(c)-c\right]dc$$

$$\Longleftrightarrow$$

$$\int_{\underline{c}}^{\bar{c}} F(c)\int_{c}^{\bar{c}} F_i(\tau|c)\,d\tau\,dc\leq\int_{\underline{c}}^{\bar{c}} F(c)F_i(c|c)\left[b_d^*(c)-c\right]dc. \qquad (4.320)$$

This is condition (4.317).

(d) Using (4.262), (4.197), and (4.143), the stated relation is valid if the following holds true

$$2\left[1-\frac{k}{d}\right]\bar{p}+2\left[\frac{2k}{d}-1\right]\int_{\underline{c}}^{\bar{c}} F(c)\left[cf(c|c)+cF_i(c|c)+\int_{c}^{\bar{c}} F_i(\tau|c)\,d\tau\right]dc$$

$$\leq 2\bar{p}-2\int_{\underline{c}}^{\bar{c}} cF(c)\left[f(c|c)+F_i(c|c)\right]dc$$

$$-2\int_{\underline{c}}^{\bar{c}} F(c)\left[\frac{kf(c|c)}{d-k}+F_i(c|c)\right]\left[b_u^*(c)-c\right]dc$$

$$\Longleftrightarrow$$

$$\int_{\underline{c}}^{\bar{c}} F(c)\left[\frac{kf(c|c)}{d-k}+F_i(c|c)\right] b_u^*(c)\,dc$$

$$\leq \frac{k}{d}\left[\bar{p}+\frac{2k-d}{d-k}\int_{\underline{c}}^{\bar{c}} cF(c)\,f(c|c)\,dc\right]$$

$$-\left[\frac{2k}{d}-1\right]\int_{\underline{c}}^{\bar{c}} F(c)\left[F_i(c|c)\,c+\int_{c}^{\bar{c}} F_i(\tau|c)\,d\tau\right] dc.$$

This is condition (4.319). □

The expected production costs are the same because all three bidding strategies prefer low-cost generators over more expensive ones. The dominance of one auction in terms of expected average prices depends on the probability density function.

Remark 4.3. *It is easily checked that all auctions have the same properties concerning expected profit, costs, and average price if marginal costs are independently draw.*

4.6.3 Case C: Full Market Capacity Utilization

The remaining demand range is covered by the following theorem.

Theorem 4.29. *Suppose $d \geq 2k$.*

(a) *Expected profits are the same for all auction types*

$$\pi_d^*(c) = \pi_u^*(c) = \pi_g^*(c)\,. \qquad (4.321)$$

(b) *Expected costs per electricity unit are equal for all auction types*

$$E[c_d] = E[c_u] = E[c_g]\,. \qquad (4.322)$$

(c) *All auction types have the same expected average price or uniform price, respectively*

$$E[p_d] = E[p_u] = E[p_g]\,. \qquad (4.323)$$

Proof.

(a) This is obvious, see (4.117), (4.207), and (4.271).

(b) This is obvious because all auction types select the production schedule with the lowest cost.

(c) This is obvious, see (4.119), (4.209), and (4.273). □

It is no surprise that none of the auction types is dominant. All market participants in this demand range are aware, that consumers may will be rationed because of a lack of production capacity for $d > k$.

Chapter 5
Conclusion

Electricity prices at power exchanges depend on some factors, e.g. production costs, supply, demand, and competitive pressure. A crucial aspect in the power market is the restricted amount of electricity a generator is able to produce. The existence of this limited production capacities influences the competitive situation strongly. Auction rules and individual production costs are further influences often analyzed.

This work analyzes the bidding behavior of two risk-neutral generators under three auction types. Each of the generators owns one power plant with the same production capacity. Producing more electricity units than the production capacity is not feasible. A well-known example for such a duopoly market is the wholesale market for electricity in England and Wales after the deregulation in 1990. The decision for a single power market based on an auction allows the operator of the electricity grid who acts also as the auctioneer to provide a high level of safety and efficiency.

The auctioneer acts as a proxy of the consumers whose demand is price inelastic. Similar to other auctions, the auctioneer sets a reserve price, which is in this work common knowledge. The highest profit a generator can earn is therefore determined by the difference between the reserve price and production costs and the lowest value of demand or production capacity. The competitive pressure is the highest if the loser, who offered the highest bid, does not produce anything. This makes it more important for a generator to win the auction at the expense of profit. Using loser's amount of sold electricity as a criterion for the competitive situation, three demand ranges can be distinguished independently of the implemented auction type.

The first range is determined by a demand, which does not exceed the production capacity of one generator. It allows only the winner to earn money. The second range is characterized by a demand higher than the production capacity of one generator but lower than the production capacity of both generators. Now the loser sells some electricity but does not fully utilize his production capacity. The third range is reached if demand equals or is higher than the production capacity of both generators. Consumers have then to accept that their demand may be rationed.

The three analyzed auctions are the uniform-price auction, the discriminatory auction, and the generalized second-price auction. The uniform-price auction is

S. Schöne, *Auctions in the Electricity Market*, Lecture Notes in Economics and Mathematical Systems 617,

commonly used because every generator gets the same price per electricity unit and no one is discriminated. The discriminatory auction is rarely used. Its payment rule stays often in conflict with the understanding of a fair price because each bidder is paid his own bid. The last auction, the generalized second-price auction, is analyzed because it leads to bids, which reflect the true marginal costs. This is economically interesting because all market participants can calculate the production costs correctly.

The competing generators have constant marginal costs. They are the realization of a known probability function. Costs are presumed to be affiliated. This allows analyzing markets with correlated production costs as well as with independent costs. All aspects of the game are common knowledge except the realized values of the marginal costs. The bids given to the auctioneer are sealed. It reflects the commonly used form at exchanges.

The work presents along with the theoretical results an example for illustration. This example is based on a probability distribution, which allows adjusting the degree of affiliation of marginal costs by one variable. Positive values stand for strong affiliated, i.e. positive correlated costs. Independent costs result if the variable is zero.

The economic analysis starts with the search of non-cooperative bidding strategies maximizing the expected profit of each generator. Unique Bayes–Nash equilibria are always found except for the third demand range in the generalized second-price auction. In such a case multiple equilibria occur. But they do not alter the results of the other key figures. The optimal bid functions of the uniform-price and the discriminatory auction given the second demand range exist only under a condition. It states that the reserve price must lie below a critical value, which depends on the probability distribution and the demand. The auctioneer has to respect this before she announces the reserve price to enforce bids calculated according to the optimal bid functions.

Bid shading exists in the discriminatory and the uniform-price auction. The generalized second-price auction enforces bids equal to marginal costs, which is an interesting economic feature. Therefore, bids in the generalized second-price auction are the lowest for all demands below the aggregated production capacity of both generators.

All auctions are cost efficient, i.e. they have the lowest possible production costs. This is a result of the selection rule in all auctions and the bidding behavior. The optimal bids increase with marginal costs. The auctioneer, who chooses the bidder with the lowest bid first, selects therefore also the generator with the lowest production costs first. If demand exceeds the production capacity of one generator, the average costs increase with the demand until both power plants are fully utilized.

Differences between the uniform-price auction and the discriminatory auction concerning the bidding behavior occur only if the loser sells something but does not fully utilizes his production capacity. While bids in the discriminatory auction smoothly increase with demand until the reserve price is reached, bids in the uniform price auction drop abruptly when demand crosses the production capacity of a single generator. The jump is a result of the shifted incentive of the generators. Because the winner gets the bid of his competitor paid he is more concerned about becoming the winner. Hence, he reduces the bid to increase his probability of winning.

Three auctions are analyzed in this work. The decision which auction type will be implemented in an electricity market often depends on the persons who have the most power. Hence, it can either be the generators or the consumers. The key figures of the generators are expected profit and the expected contribution margin. Expressions for all auctions and all demand ranges are given. The expected profit increases with demand. The highest value is reached when both generators fully utilize their production capacity. This results shows that the generators have an incentive to stay in the market and therefore, that consumers must not fear a market failure because of missing generators.

If demand is sufficient such that both generators fully utilize their production capacities or if marginal costs are independently distributed, generators are indifferent between the auctions. The preference for a specific auction depends for all other cases on the probability distribution and the reserve price. It may happen that not all generators agree on the same auction. The example presents such a case in Figure 4.36.

The preferred point of view concerning the decision which auction should be implemented is the position of the consumers. They want to pay the lowest average price per electricity unit. Because they do not know the realized production costs in advance, they use the expected average price. Expressions are given for all auctions and demand ranges. The highest expected price is reached when both generators earn the highest expected profit. Similar to the generators, consumers are indifferent between the auctions given independently distributed marginal costs or a demand equal or higher than the production capacity of both generators. Otherwise, the ranking of the auctions depends on the probability distribution and the reserve price.

This work shows that all three auction types create markets, which solve the allocation problem economically taking into account technical aspects of the electricity production. The decision of the auctioneer about the reserve price is in some cases more important than in others because it decides whether generators act according to the optimal bidding strategy or not.

Nevertheless, further research is necessary. A first extension is to analyze a market with more than two bidders. It could answer the question how many generators are necessary in an electricity market to keep the expected average price at an acceptable low level for the consumers.

Another extension is to allow generators to choose their maximal amount of electricity for sale endogenously. It is a more realistic assumption because, e.g. a generator may has technical problems to fully utilize his capacity. Another example is an electricity market divided into a wholesale market with a power exchange and a resale market with direct access to consumers. A generator would sell only free capacity at the exchange. Besides the bid price, the free decision about the production capacity for sale can lead to more strategic behavior of the generators. It may support empirical findings for the electricity market in England and Wales.[1]

The suggested extensions presumed symmetric generators. A real market consists normally of generators with different production capacities. This is especially important when a monopolistic market should be broken up. With the analysis of various market shares, it would, e.g. be possible to measure the benefit for the consumers of a deregulation. The results could be used to decide about the sizes of the new companies if the monopolist is split up. The analysis can also be used to figure out if the prices are manipulated, e.g. because of cooperating behavior among generators.

A remaining problem of the explained extensions is their static character. Auctions in the electricity market would take place daily to keep the economy running. Because the building of new power plants is time consuming, the competing generators know each other for a long time and learn from each other. They want to maximize their expected discounted profit. This could lead to strategic behavior, which is unfavorable for the consumers but may have to be accepted as the price of a working market.

[1] See, e.g. Wolfram (1998).

References

1. ap (2007) Neue Welle von Strompreiserhöhungen. Financial Times Deutschland (Online) (Jul 23). URL http://www.ftd.de/politik/deutschland/229989.html?mode=print. Accessed Sep. 14, 2007
2. Ausubel L M, Cramton P C (2002) Demand Reduction and Inefficiency in Multi-Unit Auctions. Working paper, Jul 12. University of Maryland. URL http://www.ausubel.com/auction-papers/demand-reduction-r.pdf. Accessed Jul. 22, 2003
3. Baldick R, Grant R, Kahn E P (2004) Theory and Application of Linear Supply Function Equilibrium in Electricity Markets. Journal of Regulatory Economics 25(2):143–167
4. Baldick R, Hogan W (2002) Capacity Constrained Supply Function Equilibrium Models of Electricity Markets: Stability, Non-decreasing constraints, and Function Space Iterations. POWER Working Paper, no 089, Apr 23. University of California, Energy Institute
5. BDEW Bundesverband der Energie- und Wasserwirtschaft e. V. (BDEW) (2008) Erdgas und Strom 2007: Höhere Energieeffizienz dämpft Energieverbrauch. Press release, Feb 27. Berlin
6. Beharrell S R (1991) The Electricity Pool in England and Wales: The Generator's Perspective. Journal of Energy & Natural Resources Law 9(1):45–65
7. Bikhchandani S (1999) Auctions of Heterogeneous Objects. Games and Economic Behavior 26(2):193–220
8. Bohn R E, Caramanis M C, Schweppe F C (1984) Optimal pricing in electrical networks over space and time. Rand Journal of Economics 15(3):360–376
9. Bolle F (1992) Supply function equilibria and the danger of tacit collusion: The case of spot markets for electricity. Energy Economics 14(2):94–102
10. Borenstein S (2002) The Trouble With Electricity Markets: Understanding California's Restructuring Disaster. Journal of Economic Perspectives 16(1):191–211
11. Bower J, Bunn D W (2001) Experimental analysis of the efficiency of uniform-price versus discriminatory auctions in the England and Wales electricity market. Journal of Economic Dynamics and Control 25(3-4):561–592
12. Brimmer A F (1962) Price Determination in the United States Treasury Bill Market. The Review of Economics and Statistics 44(2):178–183
13. Brunekreeft G (2001) A multiple-unit, multiple-period auction in the British electricity spot market. Energy Economics 23(1):99–118
14. Brunekreeft G, Keller K (2000) The electricity supply industry in Germany market power or power of the market? Utilities Policy 9(1):15–29
15. Bundesgesetzblatt (1998) Gesetz zur Neuregelung des Energiewirtschaftsrechts. Vom 24. Arpil 1998. Bundesgesetzblatt, Teil I 1998(23, Apr 28):730–736
16. Chao H-P, Peck S (1996) A Market Mechanism For Electric Power Transmission. Journal of Regulatory Economics 10(1):25–59
17. Crampes C, Creti A (2003) Capacity Competition in Electricity Markets. Working paper, Dec. Université de Toulouse. URL http://www.laser.univ-montp1.fr/Seminaires/crampes-creti-12_03.pdf. Accessed Jan. 14, 2008

18. Cramton P C (2003) Electricity Market Design: The Good, the Bad, and the Ugly. In: Proceedings of the Hawaii International Conference on System Sciences, Jan. URL http://csdl.computer.org/comp/proceedings/hicss/2003/1874/02/187420054b.pdf. Accessed Jan. 28, 2008
19. Crawford G S, Crespo J, Tauchen H (2007) Bidding asymmetries in multi-unit auctions: Implications of bid function equilibria in the British spot market for electricity. International Journal of Industrial Organization 25(6):1233–1268
20. dpa (2007) Eon-Chef prognostiziert steigende Energiepreise. Financial Times Deutschland (Online) (Sep 14). URL http://www.ftd.de/unternehmen/industrie/252943.html?mode=print. Accessed Sep. 14, 2007
21. Eidgenössisches Departement für Umwelt, Verkehr, Energie und Kommunikation UVEK (Eidgen. Dep. UVEK), Bundesamt für Energie BFE (2007) Überblick über den Energieverbrauch der Schweiz im Jahr 2006. Auszug aus der Schweizerischen Gesamtenergiestatistik 2006, Jun. Ittigen
22. Elmaghraby W J (2005) Multi-unit auctions with complementarities: Issues of efficiency in electricity auctions. European Journal of Operational Research 166(2):430–448
23. Energiewirtschaftsgesetz (EnWG) (2005) Gesetz über die Elektrizitäts- und Gasversorgung (Energiewirtschaftsgesetz - EnWG). Vom 7. Juli 2005. Bundesgesetzblatt, Teil I 2005(42, Jul 12):1970–2009
24. European Energy Exchange (EEX) AG (2000) The EEX Spot Market. The European Spot Market for Energy. Nov
25. European Energy Exchange (EEX) AG (2003) EEX triples yearly trading volume. Press release, Jan 6. Leipzig
26. European Energy Exchange (EEX) AG (2005) EEX starts Trading Zone in Austria. Press release, Mar 23. Leipzig
27. European Energy Exchange (EEX) AG (2006) EEX: Successful launch of the Swiss Spot Market for Power. Press release, Dec 11. Leipzig
28. European Energy Exchange (EEX) AG (2007a) EEX Product Information Power, document release: 001A. Jan 30. Leipzig
29. European Energy Exchange (EEX) AG (2007b) Geschäftsbericht 2006. Mar 29. Leipzig
30. European Energy Exchange (EEX) AG (2007c) EEX sees free market called into question by political sector. Political attacks emphasise emotional character instead of objectivity - Amendment of the Act against Restraints of Competition may lead to break with principle of liberalized markets. Press release, Jan 25. Leipzig
31. European Energy Exchange (EEX) AG (2007d) EEX and Powernext to co-operate. Press release, Dec 12. Berlin, Leipzig
32. European Energy Exchange (EEX) AG (2008) EEX: Considerable increase in trade volumes in 2007. Press release, Jan 9. Leipzig
33. European Energy Exchange (EEX) AG, LPX Leipzig Power Exchange (LPX) GmbH (2001) EEX and LPX merge to a new power exchange. First merger of two power exchanges in Europe. Mutual press release, Oct 26. Berlin
34. Fabra N, von der Fehr N-H M, Harbord D C (2002) Designing Electricity Auctions: Uniform, Discriminatory and Vickrey. Working paper, Nov 9. URL http://econwpa.wustl.edu:8089/eps/misc/papers/0211/0211017.pdf. Accessed Jan. 29, 2003
35. Fabra N, von der Fehr N-H M, Harbord D C (2006) Designing electricity auctions. Rand Journal of Economics 37(1):23–46
36. Federico G, Rahman D (2003) Bidding in an Electricity Pay-as-Bid Auction. Journal of Regulatory Economics 24(3):175–211
37. von der Fehr N-H M, Harbord D C (1993) Spot Market Competition in the UK Electricity Industry. The Economic Journal 103(418):531–546
38. Friedman M A (1963) Price Determination in the United States Treasury Bill Market: A Comment. The Review of Economics and Statistics 45(3):318–320
39. Fudenberg D, Tirole J (1995) Game theory, 4th edn. MIT Press, Cambridge (MA), London

40. Gilbert R J, Kahn E P, Newbery D M (1996) Introduction: International comparisons of electricity regulation. In: International Comparisons of Electricity Regulation. Gilbert R J, Kahn E P (eds). Cambridge University Press, Cambridge, pp 1–24
41. Goldstein H (1962) The Friedman Proposal for Auctioning Treasury Bills. The Journal of Political Economy 70(4):386–392
42. Green R J (1991) Reshaping the CEGB: Electricity privatization in the UK. Utilities Policy 1(3):245–254
43. Green R J (1996) Increasing Competition in the British Electricity Spot Market. The Journal of Industrial Economics 44(2):205–216
44. Green R J (1999) The electricity contract market in England and Wales. The Journal of Industrial Economics 47(1):107–124
45. Green R J, Newbery D M (1992) Competition in the British Electricity Spot Market. The Journal of Political Economy 100(5):929–953
46. Harbord D C, McCoy C (2000) Mis-Designing the U.K. Electricity Market? European Competition Law Review 21(5):258–260
47. Harsanyi J C (1967/68) Games with Incomplete Information Played by Bayesian Players. Part I–III. Management Science 14:159–182, 320–334, 486–502
48. Hogan W W (1998) Competitive Electricity Market Design: A Wholesale Primer. Dec 17. Cambridge (MA): Harvard University, John F. Kennedy School of Government. URL http://faculty-gsb.stanford.edu/wilson/E542/classfiles/HoganPrimer.pdf. Accessed Nov. 18, 1996
49. Joskow P L (1997) Restructuring, Competition and Regulatory Reform in the U.S. Electricity Sector. Journal of Economic Perspectives 11(3):119–138
50. Kaul M (2001) Kurspolitik von Aktienhändlern. Ein Finanzmarktmodell mit unvollständiger Information, 1st edn. Gabler Edition Wissenschaft, Deutscher Universitäts-Verlag, Wiesbaden
51. Klemperer P D, Meyer M A (1989) Supply Function Equilibria in Oligopoly under Uncertainty. Econometrica 57(6):1243–1277
52. Krishna V (2002) Auction theory. Academic Press, Inc., San Diego (CA)
53. LPX Leipzig Power Exchange (LPX) GmbH (2000) LPX Leipzig Power Exchange (Image booklet). Leipzig
54. manager-magazin.de (2007) Strompreise. EEX weist Vorwürfe zurück. manager-magazin.de (Mar. 13). URL http://www.manager-magazin.de/geld/artikel/0,2828,druck-471387,00.html. Accessed Mar. 6, 2008
55. Milgrom P R, Weber R J (1982) A Theory of Auctions and Competitive Bidding. Econometrica 50(5):1089–1122
56. Möhring-Hüser W, Morovic T, Pilhar R (1998) Strompreise für morgen. Feldversuch eines lastabhängigen Echtzeit-Tarifs. Energiewirtschaftliche Tagesfragen 48(5):307–403
57. Müller U (1999) Städtequartett wirbt um Strombörse. WeltOnline (Jun 4). URL http://www.welt.de/daten/1999/06/04/0604wi116711.htx?print=1. Accessed Feb. 15, 2002
58. Müller S (2001) Auctions. In: International Encyclopedia of the Social & Behavioral Sciences. Smelser N J, Baltes P B (eds). Pergamon, Oxford, pp 917–923
59. Myerson R B (1981) Optimal auction design. Mathematics of Operations Research 6(1):58–73
60. Neame P J, Philpott A B, Pritchard G (1999) Offer Stack Optimization for Price Takers in Electricity Markets. In: Proceedings of the 34th Annual Conference of the Operational Research Society of New Zealand, University of Auckland, New Zealand, pp 3–12. Dec 10–11
61. Newbery D M (1998) Competition, contracts and entry in the electricity spot market. Rand Journal of Economics 29(4):726–749
62. Newbery D M, Green R J (1996) Regulation, public ownership and privatization of the English electricity industry. In: International Comparisons of Electricity Regulation. Gilbert R J, Kahn E P (eds). Cambridge University Press, Cambridge, pp 25–81
63. Parisio L, Bosco B (2003) Market Power and the Power Market: Multi-Unit Bidding and (In)Efficiency in Electricity Auctions. International Tax and Public Finance 10(4):377–401
64. Rothkopf M H, Teisberg T J, Kahn E P (1990) Why Are Vickrey Auctions Rare? The Journal of Political Economy 98(1):94–109

65. Rudkevich A (1999) Supply Function Equilibrium in Power Markets: Learning All the Way. TCA Technical Paper, no 1299-1702, Dec 22. Cambridge (MA): Tabors Caramanis & Associates
66. Schiffer H-W (1994) Energiemarkt Bundesrepublik Deutschland, 4th edn. Verlag TÜV Rheinland, Cologne
67. Singh H (1999) Auctions for ancillary services. Decision Support Systems 24(3-4):183–191
68. Spiller P T, Martorell L V (1996) How should it be done? Electricity regulation in Argentina, Brazil, Uruguay and Chile. In: International Comparisons of Electricity Regulation. Gilbert R J, Kahn E P (eds). Cambridge University Press, Cambridge, pp 82–125
69. STATISTIK AUSTRIA Bundesanstalt Statistik Österreich (2007) Energiebilanzen 1970–2006. Dec. 14. Vienna
70. stern.de (2007a) Strompreistreiberei. EEX weist Vorwürfe zurück. stern.de (Mar. 13). URL http://www.stern.de/wirtschaft/unternehmen/maerkte/584614.html. Accessed Mar. 6, 2008
71. stern.de (2007b) Preistreiberei. Kartellamt untersucht Strommarkt. stern.de (Mar. 14). URL http://www.stern.de/wirtschaft/unternehmen/584678.html. Accessed Mar. 6, 2008
72. Supatgiat C, Zhang R Q, Birge J R (2001) Equilibrium Values in a Competitive Power Exchange Market. Computational Economics 17(1):93–121
73. Sydsæter K, Strøm A, Berck P (1999) Economists' Mathematical Manual, 3rd edn. Spinger-Verlag, Berlin et al
74. Thomas S (2006) The British Model in Britain: Failing slowly. Energy Policy 34(5):583–600
75. Verband der Elektrizitätswirtschaft - VDEW - e. V. (VDEW) (2003) VDEW-Jahresbericht 2002. Energie effizient nutzen. Apr. Berlin, Frankfurt am Main
76. Vickers J, Yarrow G (1991) The British electricity experiment. Economic Policy 6(12):187–232
77. Vickrey W (1961) Counterspeculation, auctions, and competitive sealed tenders. Journal of Finance 16(1):8–37
78. Wilson R B (1979) Auctions of Shares. Quarterly Journal of Economics 93(4):675–689
79. Wilson R B (2002) Architecture of Power Markets. Econometrica 70(4):1299–1340
80. Wolfram C D (1998) Strategic bidding in a multiunit auction: an empirical analysis of bids to supply electricity in England and Wales. Rand Journal of Economics 29(4):703–725
81. Wu F F, Varaiya P (1999) Coordinated multilateral trades for electric power networks: theory and implementation. International Journal of Electrical Power & Energy Systems 21(2):75–102

Index

C

D

E

L

M

N

O

P

R

S

Printing: Krips bv, Meppel, The Netherlands
Binding: Stürtz, Würzburg, Germany